The Lowdown on GMOs:
According to Science

COMPILED BY
FOURAT JANABI

CONTENTS

OPENING MINDS AGAINST MYTHS

BY KARL HARO VON MOGEL

Plant Geneticist

In the fall of 1999, I took my first classes as a genetics student at UC Davis. While I was figuring out what I wanted to focus my life on, news of a disaster struck. A group of activists destroyed a research field and a weather station, protesting against genetically engineered (GE) crops. What they destroyed was not genetically engineered at all, but was the graduate thesis project of a plant science student, among several basic research projects. This pattern would continue with other similar acts of vandalism and protest around the country and the World.

Jump forward a few years, after interests in plant genetics and science journalism started trying to pull me in two directions, another group of protesters descended on our land-grant university. This time the activists chained themselves to a DNA sculpture in a stairwell, claiming that they wanted to stop a project that was planning to cover the world in genetically engineered trees. When I interviewed the director of the Dendrome Project for the school paper, I learned that it was nothing like what the activists described at all – but was simply a database of gene sequences gathered from tree research, like the Genbank that helps scientists study everything from the genetics of human diseases to the genetics of crops and endangered species.

Elsewhere, GE crops were being destroyed, blocked, banned, and demonized. I struggled to figure out why these things were happening. The actions being taken, and the rhetoric used to justify the protests, vandalism and other acts didn't seem to square with the science that I was learning.

Humans have been seeking ways to improve their crops and livestock for millennia, and in the last century our knowledge of the fundamental genetic basis of these changes has expanded immensely. We continue to develop new tools to use this knowledge to improve our crops to handle the many demands we place on them to keep our society fed, clothed, and now, fueled. Genetic engineering is one of the most recent tools, which allows us to more directly alter what we have been changing almost blindly for thousands of years. It also expands the pool of genetic resources we can use for breeding beyond closely-related species to include the entire web of life.

We've seen many benefits from genetic engineering in medicine and agriculture, and to this day many hundreds of published studies demonstrate its usefulness, effectiveness, and safety. These studies also point the way to figuring out what we do not yet know about its full impact on the world. When I encountered this controversy I read what I could, attended forums, and interviewed people for the newspaper and for the local community radio station. Every time I heard a new claim, whether it was a random guy in Florida posting a video of him claiming to have an allergic reaction to GE corn, or a farmer in Canada being sued over accidental cross-pollination of GE canola onto his farm, I traced it to the source.

The allergy-stricken man did a double-blind allergy test and was not allergic to GE corn after all, and the farmer was spraying his canola with herbicide to collect GE seeds for replanting each year. I chased down the demons haunting these activists and always found a man or woman wearing a mask. They have largely disappeared from the limelight, yet, the mythologies propagated by these and other stories remain alive in the discussion today. Now they have mutated, multiplied, and have become part of people's personal and political identities. And with a new online media environment that is fragmented and filtered by those identities, clearing the air presents a daunting task for scientists and journalists alike.

What scientist, busy with research, teaching, peer reviewing manuscripts, and organizing and attending conferences, wants to wade into the furor fomented by people who will always doubt their results, when they can just keep advancing the frontiers of knowledge instead? What journalist, wherever they lie on the continuum from blogger to professional, will devote the time to sift through the sands of sensationalist claims to find single golden nuggets of truth, to only find comments on their articles asserting that they, too are ironically part of some grand conspiracy to deceive?

We can easily see the logic in going with the flow and keeping out of that spotlight. But that logic fails us when we think about our long-term future. If there is a significant benefit to be gained by using this technology for good, we will never realize that benefit if we do not have a concerted effort to educate people, and inspire them to learn more and explore the issue of genetic engineering. That is what motivated me to pursue a hybrid path of science and communication. People need to understand how they can relate to this technology, and why so many scientists devote their lives to doing research in this field. We need to have a robust discussion that talks about the real benefits and risks of using – and not using – genetic engineering in our lives, and it requires stepping out of our comfort zones.

I have painted you an ugly picture of a polarized and hopeless debate, but that is only how it appears on the surface. In truth, most people have not yet decided what they think about this technology and are still seeking knowledge to help them evaluate it. The scientific endeavor continues to open up new avenues of research and possibilities for the future. And there is a growing movement of scientists, journalists, environmentalists, farmers, business leaders, and other thinkers who want to help be stewards of this knowledge. They are venturing out into the darkness with their own candles to help us see the way and have the discussion that we needed years ago.

What you will find when you read this book are some important new voices who relate their direct experience and knowledge of the technology, and how we should analyze its potential role in our future. Some of them have been to the other side of the debate and returned to tell us how science, rationality, and our shared human values brought them back out again. If you have a stake in this debate over genetic engineering, whether for, against, or just one of the many people who *eat food*, you owe it to yourself to read on, shed some myths, and open your mind.

Cৰৎৎ

KARL'S BIO

Karl earned his Ph.D. in Plant Breeding and Plant Genetics at UW-Madison, with a minor in Life Science Communication. His dissertation was on both the genetics of sweet corn and plant genetics outreach. He currently works as a public research geneticist in Madison, WI. His favorite product might just be squash.

— *Karl Von Haro Hogel*

Twitter: @kjhvm

PART I.

Introduction

I hope that every student at one point in their life has the opportunity to have something that is at the heart of their being, something so central to their being that if they lose it they won't feel they're human anymore, to be proved wrong because that's the liberation that science provides. The realization that to assume the truth, to assume the answer before you ask the questions leads you nowhere.

Lawrence Krauss

WHO'S AFRAID OF GMOS?

BY ALAN MCHUGHEN

Molecular Biologist

GMOs (genetically modified organism) are products of technologies developed during the 1970s and 1980s that allow researchers to take DNA (i.e., genetic information) from any plant, animal or microbe and combine it with the DNA of any other plant, animal or microbe. This technique provides an ability to transfer genetically controlled traits among different species, for example to have bacteria manufacture insulin, or allow a water-thirsty crop to withstand drought.

For various reasons, this recombinant DNA technology, rDNA, is scary to some. Back in 1998, Prince Charles wrote in *The Daily Telegraph* that "I happen to believe that this kind of genetic modification takes mankind into realms that belong to God and to God alone." Still others fear the apparently uncertain safety record of GMOs, and the idea that this technology may inadvertently introduce safety hazards into foods. And another large segment fears not the technology per se but rather the idea of technology and big multinational corporations dominating the food supply. Leading GMO seed developer Monsanto, for example, is the company many people love to hate.

The reality is that GMO technologies have given us many useful products, from human insulin to safer crops grown with fewer pesticides. Moreover, in over 30 years of experience, according to authoritative sources such as the U.S. National Academies and the American Medical Association, there is not one documented case of harm to humans, animals or the environment from GM products. That is an impressive track record, considering the extent of GM products in pharmaceuticals, agriculture, food, and industrial applications.

So why are so many still fearful of this technology? One simple answer is junk science, and its carefully crafted use as a weapon of mass fear.

Jeremy Rifkin was the first junk dealer to make big money by scaring people about the potential dangers of genetic engineering. Rifkin is no scientist, but an economist and prolific story spinner — the author of numerous books such as *Algeny* (1983) and *The Biotech Century* (1999). They are all, apparently, classified as non-fiction. None is peer reviewed, however. The late evolutionary scientist Stephen Jay Gould referred to *Algeny* as "a cleverly constructed tract of anti-intellectual propaganda masquerading as scholarship," and in 1989, *Time* magazine ran a story titled "The Most Hated Man in Science."

But biotechnology is not Rifkin's main target. His real bugbears are capitalism and modern agriculture; the hybrid progeny of these two foretell, according to Rifkin's junk-science theory, the demise of humanity.

Greenpeace and other special interest groups, such as Friends of the Earth and the U.K.'s Soil Association, have deployed their considerable media-manipulating machinery to spread more scare stories.

Activists claimed they were performing a public service by alerting locals in Africa that GM foods from the United States would render the men impotent. In the Philippines, people were told, and some convinced, by activist scaremongers that merely walking through a field of genetically modified corn could turn heterosexual, virile men gay. European activists went to Zambia during the height of the 2002 famine and convinced then president Levy Mwanawasa that the GM corn in food aid contributed by the United States was "poison." As reported by the British Broadcasting Corporation, Mwanawasa duly locked up the food in the warehouses — the same GM corn eaten without incident by millions of Americans and others around the world — and then watched his subjects die, insisting that such a fate was preferable to eating "poison." That is, until the starving Zambians broke into the warehouses and gorged themselves healthy on the allegedly poisonous corn.

Another popular junk scientist is Jeffrey Smith, who has penned several books decrying the alleged hazards of modern agriculture, saving his most potent venom for genetically modified crops and foods. Smith's self-published, non-peer-reviewed Genetic Roulette, for example, expounds upon already questionable reports — almost all from non-peer-reviewed sources — in a confident, technical tone that suggests that he actually has some scientific or medical credentials. Indeed, his acolytes and sycophants on the internet have taken to referring to him as "Dr." Smith, providing him a mantle of scientific and medical credibility. But closer inspection of Smith's CV reveals that the closest he has come to scientific credentials is working as a ballroom dance instructor and a flying carpet yogi.

Social media fuel the fire: Anyone can publish any outlandish junk science claim on the Internet. But when a plant breeder develops a strain of rice that is enhanced to help overcome vitamin A deficiency, rampant in poor tropical countries, the media interview (and give prominence to) pseudoscientific scaremongers such as Smith instead of authentic experts in nutrition or agronomy, people who might actually bring legitimate questions and concerns to the discussion.

In 2012, French scientist Gilles-Éric Séralini and his team published a peer-reviewed paper (since retracted) that claimed harm to test animals after they were fed GM corn for two years. Séralini boasted that his paper was the first long-term GM feeding trial. But Séralini, and later his disciples, failed to note the many other peer-reviewed, long-term GM feeding studies, including one in the journal in which his claims appeared, that concluded the opposite about the effect of GM food on animals: that such food was as safe, or safer, than regular non-GM food and feed.

Fortunately, there are reputable sources out there, too. Food safety and environmental sustainability are essentially scientific subjects, so we must reply on legitimate, credentialed scientific and medical expertise to assess and evaluate risks to the food supply or sustainability. Such professional bodies — non-corporate entities serving the public good — as the US National Academies of Science, US Institute of Medicine, UK's Royal Society, American Medical Association, French Academy of Science, American Dietetic Association, Third World Academy of Sciences, along with government agencies such as USDA, FDA and EPA in the US, Canada's CFIA and health Canada and others have all studied the risks associated with GM technology, and invariably conclude that the risks are no greater (and sometimes less) than with conventional breeding methods.

In contrast, I could find no professional scientific or medical bodies anywhere in the world publishing a peer reviewed study concluding GM methods were higher risk than conventional breeding methods.

But the sources suffer from relatively low conventional and social media profiles: They tend to appear near the bottom of Internet searches, even though they rank at the top of scientific credibility. They are mainly the professional scientific and medical associations, groups such as the U.S. National Academy of Sciences, the British Royal Society and the American Medical Association. These groups are not selling GMOs, and therefore are immune to the charge often leveled by pseudo-scientists and anti-technology activists that the private sector lies, cheats and steals to show its products in a good light.

When it comes to the safety and sustainability of GM technologies in agriculture and food production, the U.S. National Academies of Science have conducted expert reviews of GMO safety going back to 1986. All are freely available online, if one knows where to look. Every single one of these studies has reached the same general conclusion: GMOs are no more hazardous than are other forms of breeding. A major investigation in 2004 into the safety of genetically engineered foods, for instance, concluded that GM technology is not inherently hazardous: "To date, no adverse health effects attributed to genetic engineering have been documented in the human population."

A more recent study, from 2010, investigated the impact of genetically engineered crops on farm sustainability in the United States. This study concluded that genetic engineering technology has produced substantial net environmental and economic benefits compared with the use of non-GM crops.

Similar studies also are conducted by scientists in other countries around the world. That includes the last bastion of backward thinking against agricultural GMOs, the European Union. There, anti-science advocacy groups have been successful in scaring much of the public. To support the European political leadership that has sought scientific justification for banning

GMOs, the European Commission has been a major sponsor of public research into the safety of GMOs for over 25 years. Unfortunately for the European politicians who'd hoped to reveal some new hazards (and thus justify a ban on GMOs), all of the EU-funded research to date concludes the same as all other public studies into the safety of GMOs: that GM technology poses no new risks.

In 2001, the EU scientific community issued a report summarizing its research findings: Eighty-one research projects into GMO safety conducted by 400 teams of public scientists in non-commercial labs at a cost of 70-million euros concluded that GMOs are no more hazardous than are other forms of plant breeding. A follow-up report published in 2010 reflected the same theme, documenting 50 additional GMO safety projects funded by EU taxpayers and involving more than 400 public, non-commercial labs at a cost of more than 200-million euros. Their conclusion: GMOs are no more hazardous than other forms of breeding.

Read on to find out why.

൬ൈ

ALAN'S BIO

Alan McHughen is a public sector educator, scientist and consumer advocate. After earning his doctorate at Oxford University, Dr. McHughen worked at Yale University and the University of Saskatchewan before joining the University of California, Riverside. A molecular geneticist with an interest in crop improvement and environmental sustainability, he helped develop US and Canadian regulations governing the safety of genetically engineered crops and foods. He served on US National Academy of Sciences panels investigating the environmental effects of transgenic plants, a second investigating the safety of genetically engineered foods and helped review a third looking at sustainability and economic impacts of biotechnology on US agriculture. Having developed internationally approved commercial crop varieties using both conventional breeding and genetic engineering techniques, he has firsthand experience with the relevant technical, biosafety and policy issues from both sides of the regulatory process. As an educator and consumer advocate, he helps non-scientists understand the environmental and health impacts of both modern and traditional methods of food production. His award winning book, 'Pandora's Picnic Basket; The Potential and Hazards of Genetically Modified Foods' uses understandable, consumer-friendly language to explode the myths and explore the genuine risks of genetic modification (GM) technology. More recently, Dr McHughen served as a Jefferson Science Fellow at US Department of State and as a Senior Policy Analyst at the White House.

This chapter is based on Alan's article for the c2c journal: *Who's Afraid of the Big Bad GMO?*

— *Alan McHughen*

PART II.

Beware of false knowledge; it is more dangerous than ignorance.

George Bernard Shaw

Q&A: 21 Questions

BY FOURAT JANABI

I was once anti-GMO. At the time, I lived in Europe where being anti-GMO was simply the default position; I never felt the need to impinge upon others my anti-GMOness. Still, I was once anti-GMO. Even if I was the only one, I'd still be ashamed.

Ashamed is a strong word. Why ashamed? Because I never listened to the experts; because my knowledge of biology, which I'd assumed to be fairly solid, was riddled with more holes than Swiss cheese; because I thought I knew what people wanted more than they did. And worst of all, because I thought I could answer that last question for them...I should have realized that the choice cannot be mine to choose: that of using GMO seeds that required fewer inputs than non-GMO seeds, but my ignorance compelled me to speak, even to raise my voice on occasion; never, however, to listen to the evidence.

The following three Q&As in this chapter are the culmination of the journey I took from being anti-GMO, where I didn't understand a thing about molecular biology, to being pro-GMO, not because I stood to gain anything, but because I started to understand some of the science behind it.

These Q&As will take you on a journey to a rarer side in the kerfuffle over GMOs reverberating throughout Europe, America, and much of the rest of the rich world. The following three interviews: with a plant geneticist; a family farmer; and a CEO of a biotechnology firm will give you access to rarely explored viewpoints in the media, where instead of getting expert opinions, you hear from those who've never experimented, planted, nor starved pretend to know everything about the science of molecular biology, agriculture, and the state of the world.

Opinions are a dime a dozen. What matters is experiment, evidence and experience, and in the following three interviews, and the articles thereafter, you'll find plenty of all three and a good deal more. I hope you enjoy the read.

— Fourat Janabi

Twitter: @Fouratj

WITH THE MOLECULAR BIOLOGIST

Q&A with Kevin Folta

— —

Q1.

What is the main thing — or is it general — about GMOs that the public routinely confuse, or get wrong, when discussing and debating their impact?

There are so many misconceptions. The first is a fundamental one, that being that there is a debate at all. There is no debate among scientists in the discipline of plant molecular biology and crop science. Sure you can find someone here and there that disagrees, but there is no active debate in the literature driven by data. There are no hard reproducible data that indicate that transgenics (GMOs) are dangerous or more potentially dangerous than traditionally bred plant products.

If I had to nail down the most annoying misconceptions, they would include that all scientists are just dupes of big multinational ag companies. Anyone that presents the consensus of scientific interpretation of the literature is immediately discounted as some corporate pawn. There's nothing further from the truth. Most of us are hanging on by a thread in the days of dwindling federal, state and local support for research. The attacks on the credibility of good scientists hurts our chances to stay in academic labs and consider the cushy salaries and job security with the Big Ag corporate monstrosities we chose not to work for when we took jobs working for the public good. That's pretty sad.

There is this allegation that we hide data or don't publish work that is inconsistent with corporate desires. They need to get one thing straight. We're not in the public sector because we are excited about listening to some corporate mandates. No thanks. We're here for scientific freedom and to discover the exceptions to the rules and define new paradigms.

If my lab had a slight hint that GMOs were dangerous, I'd do my best to repeat that study, get a collaborator to repeat it independently, and then publish the data on the covers of Science, Nature and every news outlet that would take it. It would rock the world. Showing that 70-some percent of our food was poisonous? That would be a HUGE story — we're talking Nobel Prize and free Amy's Organic Pot Pies for life! Finding the rule breakers is what we're in it for, but to break rules takes massive, rigorous data. So far, we don't even have a good thread of evidence to start with.

The other huge misconception is that you can "prove something is safe." Nothing can be proven safe. We can only test a hypothesis and show no evidence of harm. You can't test all variables — nobody could. We can ask if there is a plausible mechanism for harm. If there is, we can test it. If there isn't, we can do broad survey studies. A scientist can search for evidence of harm — a scientist can never prove something is safe.

Q2.

In what ways might GMOs be most beneficial to our biosphere, and why might organic farming not be as good as to get us there?

There is no doubt that transgenic plants can be designed to limit pest damage with lower pesticide applications. That is well documented by the National Academies of Science, the best unbiased brains in our nation. Most data is for cotton and maize, and show substantial reductions (like 60%). Transgenic potatoes were amazingly successful in Romania until they joined the EU and had to go back to insecticide-intensive agriculture. Even glyphosate resistance traits, for all of their drawbacks in creating new resistant weeds, replace toxic alternatives.

Conventional farming takes fuel, labor, fungicides, pesticides, nematicides and many other inputs. Water and fertilizer are in there too. There are genes out there in the literature that address most of these issues. Scientists in academic labs discover these genes and define their function in lab-based GMOs that never are used outside the lab. The regulatory hoops are too difficult and expensive. Only the big companies can play in that space. Even little companies like Okanagan Specialty Fruits have to deal with the nonsense from those that hate the technology. Opposition to the science keeps the big guys in business, because nobody else can compete.

Who loses? The farmer, the consumer, the environment, the academic scientist and most of all the people around the world that don't get enough food and nutrition. Who gains? Big Ag.

Q3.

What do you consider the most important aspect of differentiating the good from the bad when it comes to considering science? i.e., What is the first thing you look for after reading a study?

In the short-term I consider the system studied. Was it an animal system or cells in a dish? Most of the anti-GMO work is done on cells, especially cell lines that sound scary (like ovary, testis or fetal cells) but have little relevance to the complexities of animal systems. If done in animals, was the experiment properly controlled? Do the researchers SHOW the controls (like they conveniently omitted from Seralini's 2012 rat-cancer work in Figure 3). Many studies that look good compare a GMO to an unrelated plant type. It is just not a valid comparison. Plants produce toxins and allergens, so you need to test the same exact plant without the added gene. If they do the rest of this properly then they need to run sufficient numbers and use good, common statistics. If they do all of this the work is publishable after peer review and should go into a decent journal, not some low-impact journal that publishes incomplete work or work that over steps the data.

A lot of junk escapes peer review. Reviewers and editors are overstressed and overburdened these days. We do the work as service for the field. Occasionally a paper slips by in a lower-impact journal. You'll find most of the anti-GMO papers there.

Another important attribute of good work is demonstrating a mechanism. For instance, just don't tell me that you found some evidence of GMO harming cells. Tell me how. How does it happen? If the phenomenon is real the mechanism should be dissected out in a year's time. Omics tools are incredibly sensitive and we can detect small differences in gene expression and metabolic profiles. If GMO harm was real, the authors would define that mechanism, then collect their Nobel Prize and Amy's Pot Pies.

The ultimate test is reproducibility. You'll see that the best "evidence" for harm from GMOs comes from obscure journals, aging references that were published and heavily refuted by the scientific community (Puztasi, Seralini, etc), and work that was never repeated by outside labs. These are flash-in-the-pan works that never are expanded beyond the seminal study. The best sign of real science, good science, in an edgy area is that it grows. You see more scientists pile on, more research, more funding and bigger ideas. Models expand, mechanisms grow.

That just does not happen in the anti-GMO literature. The same authors publish a paper and then it goes on the anti-GMO websites and gains attention — while it dies in the scientific literature with no follow-up.

Q4.

Is there any split in the scientific community as to the safety of GMOs? If so, where does the split lay?

There are splits in the scientific community like there are splits for climate change and evolution. You have scientists like NIH Director Francis Collins that support creationist leanings. You have a small set of meteorologists and atmosphere scientists that claim that climate change is not real. There's always room for a dissenting opinion out there, but they usually don't have good evidence, just belief.

The same is true in biology and plant science. There are a few out there that let philosophy rule over evidence, but they are not at the edge of research. In the circles I work with there is consensus about the safety and efficacy of the technology. Even those that study organic and other low-input production systems support biotech as a way to do their jobs even better. That's a strange relationship many don't expect. You'll not see anti-GMO writing from too many tenure-track scientists at leading universities. There is confusion on this. The Union of Concerned Scientists is frequently used as evidence that scientists are against this technology. When you read who they are and what they do, they are activists. They don't do research or publish in the area of biotech. There are also others that claim to be experts or exploit some tenuous university affiliation to gain credibility. They should be looked at as deceitful, but they are accepted and believed with great credibility. People like Joe Mercola, Jeffrey Smith and others sure sound like they know what they are talking about but they are not experts. Even Charles Benbrook, a guy with a great career and a wonderful CV, goes off the deep end on the topic.

Readers need to apply all of the filters discussed here. What the data really say, who did the work, and if it was reproduced independently are the most important criteria in separating reality from fiction in the GMO topic.

Q5.

Are there any verified negative aspects of using GMOs?

There are some drawbacks. The first is the well-described selection for resistant weeds when the glyphosate resistance trait is used. The first problem is that a few tough weeds are just resistant to glyphosate. They either don't absorb it, they break it down, shove it into a place where it can't function, or they have an EPSPS enzyme that is unaffected. When the rest of the weeds in an area are removed with glyphosate, these other ones (like bindweed, velvetleaf) flourish without competition. They are resistant, they always were resistant, and they require higher orders of control such as alternative chemicals or higher doses.

The big issue is evolution. When grown over massive acreage there is always a chance of a mutation leading to changes that allow the plant to survive glyphosate treatment. A plant that was not resistant, now is, and it can spread rapidly without competition. These cases require higher doses of herbicide and new technologies will use additional herbicides to combat them. We always should be trying to minimize our treatments and environmental impacts. "Stacking" traits like glyphosate resistance and 2,4-D resistance will work well (such as the "Enlist" formula) with low likelihood of resistance. It should have been done this way in the first place.

In the future, improved surfactants (wetting agents) and perhaps variations on the chemicals themselves may allow equivalent control with decreased environmental impacts.

The other negative aspect is resistance to the Bt proteins. The chance of this resistance is minimal, but formal. For proper control farmers need to maintain forages to ensure non-resistant pests and their predators are present, decreasing the likelihood of resistance. While the likelihood is minimal, there is resistance seen in corn root worms, probably because it is impossible to develop a forage for them. Resistance for pests and corn root worms is not a GM problem — it is a farming problem. The pests can be tough and Bt is an excellent strategy, which is why we need to keep innovating around that technology.

In all of these cases the specter of "super-weeds" and "superbugs" are way overblown. A super-weed is resistant to one herbicide, hardly super. Same for insect pests.

Another example of a problem is the effect of glyphosate on amphibians and aquatic environments. Glyphosate addition to standing water changes community structure of resident organisms. Glyphosate also has minor developmental effects on tadpoles, most notably an increase in their tail length associated with the presence of predators. While small differences it is important to consider them and keep studying them. It also is important to keep in mind that these effects are minor compared to the effects of other agricultural chemicals like atrazine or chemicals used in organic cultivation like rotenone. This is really a farming problem, not so much a GM problem, but certainly a place for improvement.

Q6.

How do scientists know that a gene will play nice in its new genome home? A key talking point against GMOs is uncertainty. How certain are scientists, and what evidence justifies that certainty?

First, we know a lot about plant biology and how genes work. Before a company even attempts to test, let alone commercialize, a gene of interest, massive trials are performed to understand its function, its interaction partners, its other roles in the cell. It is possible to understand the potential risks and benefits long before the transgenic plant is even constructed. Scientists know exactly what to expect when the plant is made. It can then be tested.

Fifteen years ago we could claim ignorance and uncertainty. We could not be certain where a gene landed, if it disrupted another gene or process, or if it even landed intact. There were good ways to get sound ideas and certainly no unintended consequences were expected. When you look at this process today it is amazingly precise. While we still can't place a gene in a given spot in the genome, we can tell exactly where it did integrate. It takes $1000 and a week to sequence a genome these days, so you can map precisely where the inserted gene resides in the genome. No problem!

Moreover, it is possible to perform highly-sensitive metabolomic tests that query vast swatches of secondary metabolites. We can tell easily if non-target genes are affected. It is easy to tell if something is perturbed outside of the intended gene or process.

Throughout this discussion it is important to remember that genomes are dynamic, changing, mutating and expanding all the time. Natural mobile DNAs, viruses, other materials bang around a genome and create new changes that lead to variation in a species. Compared to what is there naturally the added T-DNA is just a drop in the molecular bucket.

To summarize, from the whiteboard to the plant scientists know what a gene does, where it integrated and what it affects with extraordinary resolution. The precision is amazing, and getting amazing-er. The greatest uncertainty comes from non-GM alterations like mobile DNAs that constantly change plant genomes in ways that are unpredictable and not easily traced.

෬৪

KEVIN'S BIO.

Kevin Folta is the Associate Professor and Interim Chairman of the Horticultural Science department at the University of Florida. He has published two books on genomics: *Genetics and Genomics* (2009) and *In Genetics and Genomics of Crop Plants* (2011), while also co-authoring over 70 scientific papers in respected journals.

A prominent science communicator, you'll find him all around the web trying to set the science straight on GMOs, and is often seen handing out his personal email address to complete strangers to address the concerns of the public. He frequently writes on GMOs on his website, Illumination. In other words, he is a shill for science. In June of 2013, he participated in a Forum hosted by the CATO institute to discuss Biotechnologies role in future food production. (Leading anti-GMO activists, Jeffrey Smith and Gilles Seralini bowed out of the debate at his inclusion.)

He has a Ph.D. In Molecular Biology from the University of Illinois in Chicago. He has won the Distinguished Mentor of Undergraduate Research (2007), the NSF CAREER award (2008), the LA&S Golden Alumni Award (2009) and Foundation Research Professor (2010).

- *Kevin Folta*

@KevinFolta

THE FAMILY FARMER

Q&A with Brian Scott

— —

Q1.

Why do you use GMOs?

I like to call GMO a tool in my toolbox. Biotech isn't a silver bullet for every problem, but it's still a powerful tool. We use traits like Bt and Roundup Ready (RR) on many of our acres, but not all of them. All our soybeans are generally RR, while only some of our corn carries that trait. Popcorn and wheat, our other crops, are not available in GMO varieties. Some of our corn acres are dedicated to waxy corn production, and we generally don't buy them as RR. Built-in insect resistance in Bt corn along with seed treatments mean it's a very rare event that we have to treat a crop in season for pests. That means we prevent soil compaction by keeping another piece of equipment out of the field. It also means a sprayer doesn't need to filled with water, fuel, and pesticide which is good for the earth and the wallet.

Q2.

What incentives are there for using GMOs?

There can be incentives such as buying traited crops and certain chemistry (herbicide, etc) as a bundle to receive price discounts. Some crop insurance plans also offer a biotechnology discount. I think that says a lot about the effectiveness of GMO. If an insurance company is willing to give you a discount, they must believe those crops lead to less crop insurance claims.

Q3.

As many activists allege, are you a slave to Monsanto once you sign their contract?

I'm certainly not beholden to any seed company. I can plant what I want and manage it how I see fit. Do I sign an agreement that stipulates certain things when I buy patented seeds? Yes. Do patents only apply to biotechnology? No. These agreements are not nearly as binding as people would lead you to believe. The most viewed blog post I've put online is an outline of my 2011 Monsanto Technology Use Agreement. In the post I break down the line items in my own words, but I also provide the reader with a scanned copy of the agreement pulled straight from my filing cabinet. This allows anyone to read the agreement for themselves. In short, if I buy seed from Monsanto, Pioneer, etc. nothing binds me into buying seed from them the following season. Nothing says I have to use their brand of herbicides or insecticides. Believe what you will about farmers being slaves to seed companies, but you've got to talk to a farmer before your mind is set in stone.

Q4.

Do you think you should be able to reuse the seeds you purchase from Monsanto? If not, why not?

That's a tough question. For my purposes, if I wanted to save seed it would be soybean seed. All of our corn is hybrid corn. Hybrids don't necessarily produce seed identical to the parent plant. Therefore, planting that seed the next season would give you an unknown result. Soybeans self-pollinate so they remain true to themselves genetically. If I saved seed I would need to take a little extra care and expense to clean and possibly apply seed treatments to protect young seedlings. Right now my view is that of a division of labor. Farmers are great at producing high quality and high quantities of crops. The seed companies have the know-how and resources to breed great plants. I think that's a great combination for success. I'm not saying farmers couldn't develop their own seed. Successful farmers are some of the smartest people I know, and can do anything if they choose to.

I also believe since it takes several years and millions if not billions of dollars to bring an innovative new variety to market, that any breeder large or small should be entitled to benefit financially from said variety for some period of time via a patent system.

Q5.

What is the most glaring factual error, if any, made by activists when discussing GMO seeds?

I often ask people what they think about crops that produce their own chemical defenses naturally, and I find a good number of people aren't aware that some crops do this. For example cereal rye has an ability to suppress weeds. This quality is called allelopathy. Many plants are naturally resistant to herbicides.

Think about your lawn. Spraying 2,4D on your grass to kill dandelions and other weeds won't harm your lawn. Grasses, which include corn and wheat, have a natural tolerance to that chemistry. Biotech may be allowing plants to do new things, but we are really just mimicking something nature has already shown us is possible.

I see many people say that seeds are soaked in glyphosate which is the active ingredient in Roundup. I'm not really sure where that idea comes from, but seeds are not somehow filled with herbicide. I think it's possible people are confusing herbicides and insecticides thinking Bt and Roundup are the same thing. Bt traits protect crops like corn and cotton from pests like European corn borer.

Another fallacy is that GMO crops failed in the drought of 2012. As if somehow during the worst drought since 1988 or maybe even the Dust Bowl era nature was supposed to give us a normal yield because our crops are able to protect themselves from pests and be resistant to certain herbicides. Drought tolerant varieties of corn were not widely available to growers in 2012. I've grown Pioneer's version of drought tolerant corn in a test plot. It beat everything else in the plot hands down. Wide availability of drought tolerant corn varieties will spread in the next year or two. Drought tolerance and water use efficiency could be game changers for water use in the highly irrigated areas of the Great Plains. It should also be noted that all the corn being marketed as drought tolerant was brought to fruition by conventional breeding techniques except for Monsanto's. Theirs will be the one genetically modified version.

Farmers make plans on how to plant and manage their crops several months before actual fieldwork begins. In the end we all understand that weather will be the ultimate factor in determining the success of those plans. In agriculture there are countless variables in play when managing a crop, and the one thing you have no control over is the weather. It can rain too much or not enough. Temperatures may be great for crop growth, or they may be too hot or too cold. Farmers must do all they can to realize the potential of a seed, but nature will always dictate a large portion of yield.

Q6.

Are there any practical negatives to using GMOs on your farm?

The one downside I see to biotech is what I call the free-rider problem. Technology needs to be used as intended to work properly. For example, if you have one farmer in the area not planting refuge acres like he should because he figures all his neighbors are doing it then he possibly could be placing undo selection pressure in his fields for pests controlled with BT traits. This problem is being solved now by seed companies getting approval for refuge in a bag products. Now instead of needing to plant a non-BT refuge within certain guidelines the refuge seeds are already mixed into the bag of seed the farmer buys. I think it's good for the industry to self-police before government gets involved. The same problem can be had with farmers who don't rotate their herbicide modes of action. You can't just do glyphosate year after year with nothing else in your program. It's too bad products like Dow's Enlist or Monsanto's Roundup Ready Extend which have multiple modes of action didn't come to market first. They will be out soon and they will be great products.

Q7.

Are there any theoretical negatives or detriments that you, as a farmer, must constantly keep on the lookout for?

You just have to remember Agronomy 101. Biotechnology makes farming more sustainable but let's not mistake that for less involved management plans. Roundup Ready was and still is a fantastic technology, but we are learning it may have been too much of a good thing in some. I'm glad to see industry publications taking the lead and telling farmers to get back to basics of rotating not just crops, but pesticides as well. A renewed interest in over-cropping is something I can envision being mainstream in ten years' time. Chemicals are great tools and their safe use is proven time and time again, but let us remember that there are many more ways to manage pests, weeds, and diseases. Finding the practices for all methods will only serve to better all farms.

༄༅༄

BRIAN'S BIO

Brian Scott is a fourth generation corn, soybean, popcorn, and wheat farmer from Indiana. He farms alongside his father and grandfather. Shortly after graduating from Purdue University in 2003 with a degree in Soil and Crop Management he married his wife Nicole with whom he has one son. Brian shares stories from his farm and discusses agriculture issues in his blog, *The Farmer's Life*.

- Brian Scott

@TheFarmersLife

THE BIOTECH CEO

Q&A with Neal Carter

— —

Q1.

What prompted your company to create a GM non-browning apple? Why not, for example, try to do the same with hybridization?

Our motivation for developing biotech apples, and all our other projects under development, is to introduce value-added traits that will benefit the tree-fruit industry. We have chosen to focus specifically on non-browning Arctic® apples as our flagship project for a number of reasons. One of the chief ones is that apple consumption has been flat-to-declining for the past two decades and we are confident the non-browning apple trait can create a consumption trigger while also reducing food waste throughout the supply chain.

Another key motivation is ever-increasing demand for convenience. Arctic apples are ideally suited for the fresh-cut market, which is expensive to enter because of the browning issue. We often refer to the consumption trigger that convenient "baby" carrots created – they now make up 2/3rds of all U.S. carrot sales!

As for why we use biotechnology to achieve this, it's because we knew we could make a comparatively minor change safely, relatively quickly, and precisely. We silence only four genes, specifically, the ones that produce polyphenol oxidase, which is the enzyme that drives the browning process. We do so primarily through the use of other apple genes, and no new proteins are created. If we were to attempt to breed this trait conventionally, we could easily spend decades trying with no guarantee of success.

Q2.

What benefits will the Arctic apple bring to the food market? Are there quantitative studies that can predict how effective it could be?

In addition to addressing stagnant apple consumption and tapping into the underutilized fresh-cut and foodservice markets, Arctic apples offer plenty of other benefits throughout the supply chain.

For growers and packers, non-browning apples can help significantly reduce the huge number of apples that never make it to market because of minor superficial marks such as finger bruising and bin rubs. So much of the food produced today is wasted purely for cosmetic reasons. This extends to retail where the non-browning trait can have a big impact on shrinkage and making displays more attractive while also offering exciting new value-added apple products.

Consumers will also benefit from throwing away far less fruit at home – how many apples get bruised up on the way back from the grocery store or in kids' lunch-boxes? Our goal is helping consumers, especially kids, eat healthier and waste less food. Last year, one grade 2 teacher wrote about how excited she is for non-browning apples, explaining that she sees countless perfectly good apples and apple slices thrown out by her students due to minor browning and bruising. Consumers will also enjoy other tangible benefits like new opportunities for cut apples in many cooking applications.

As for quantifiable evidence showing the value of these benefits, food waste has been a major issue over the past year with recent estimates from the UN's Food and Agriculture Organization suggesting around one-third of food produced is wasted. The numbers are even worse for fruit, where around *half* of what's produced never ends up getting eaten.

As far as the potential to create a consumption trigger, the produce industry is full of examples of how making fruit more convenient, especially for the foodservice industry, results in huge consumption boosts. We mentioned how baby carrots now make up two-thirds of carrot sales and reports tracking major fruit and vegetable consumption trends frequently emphasize convenience. One example explains one of the most prominent, ongoing trends "is a consumer demand for foods of high and predictable quality that offer convenience and variety." Arctic apples satisfy all these requirements.

For apples, specifically, there's lots of attention given to how various chemical treatments can slow browning and plenty of attempts to conventionally breed low browning varieties (though this is quite different from being truly non-browning). For instance, a notable 2009 publication from the *Journal of Food Engineering* discusses how "the market for fresh-cut apples is projected to continue to grow as consumers demand fresh, convenient and nutritious snacks". Yet it also explains that the "industry is still hampered by-product quality deterioration" because when "the cut surface turns brown; it reduces not only the visual quality but also results in undesirable changes in flavor and loss of nutrients, due to enzymatic browning." Again, Arctic apples address these issues.

Finally, some of the most convincing evidence that the non-browning traits will provide substantial value both apple producers and consumers have told us so! In 2006/07 we surveyed a number of apple industry executives, 76% of whom told us they were interested in Arctic apples. In focus groups, we have found that over 80% are positively interested in Arctic apples and 100% of participants wanted to try them. Even more encouraging, when we surveyed 1,000 self-identified apple eaters in 2011, we found that their likelihood to buy Arctic apples continued to increase the more they learned about the science behind them!

Q3.

How many, and how intensive, were the studies performed to show Arctic apples are as safe as other apples? Were the studies peer-reviewed? If so, by whom?

Before getting into the specifics, it's important to put things in perspective to show how rigorous the review truly is; particularly arduous for a small, resource-tight company like ours.

So Arctic apples, our very first project, still haven't been commercialized 17 years after we were founded and over a decade after we proved the technology and planted them! That means we now have over ten years of real-world evidence that Arctic trees grow, respond to pest and disease pressure, flower, and fruit just as conventional trees do.

Over this time, our apples have likely become one of the most tested fruits in existence. This makes detailing all of the specific tests impossible here, but we encourage anyone interested to view our extensive, 163-page petition on the USDA's website, which provides full details.

Quickly highlighting some of the key ones:

- Trees were closely monitored by a third-party horticultural consultant for any difference in their response to pests

- Agronomic data including how fast trees grow, how much fruit they produce, etc. was recorded by a third-party

- Experiments were completed to monitor pollen spread and potential for cross-pollination, resulting in two peer-reviewed papers

- Nutrition and composition of mature fruit was tested and deemed equivalent to controls

- Possible presence of novel proteins tested and confirmed none present

These tests were performed by a variety of reputable groups and individuals, some third-party, some in-house. Our field trials were monitored and data was collected by independent horticultural consultants and an Integrated Pest Management specialist.

Of particular importance is the fact that there are no proteins in Arctic fruit that aren't in all apples. This shows there's nothing "new" in our apples that will affect consumers. This is expected as we silence the genes that cause browning, rather than introduce new attributes. To give an idea of how sophisticated the tests used to prove this are, they would be able to detect a single penny amongst 100-250 ton coal-sized rail cars! We are confident Arctic apples are safe, and soon, we anticipate FDA's confirmation of this.

So what has all of this extensive testing taught us? Exactly what we thought it would – Arctic trees and fruits are just the same as their conventional counterparts until you bite, slice or bruise the fruit!

Q4.

Can you name a few of the misconceptions – if any – that people associate your company with, or accuse your company of, when they find out you're a biotech company? If there are misconceptions, why are they wrong or miss the big picture?

Absolutely – just as there are countless misconceptions about biotech foods in general, there are also plenty of myths about our company and Arctic apples. In fact, one of our most popular blog posts ever is titled "Addressing common misconceptions of Arctic orchards and fruit".

We invite readers to visit that post and explore our site in general for more details, but the two most common misconceptions about Arctic apples are:

1. Arctic apples will cross-pollinate with other orchards, causing organic orchards to lose organic certification: No organic crop has ever been decertified from inadvertent pollen gene flow. Even if pollen from an Arctic flower did pollinate an organic or conventional fruit, the resulting fruit is the same as the mother flower.... not that of the pollen donor. Additionally, we are implementing numerous stewardship standards to ensure cross-pollination won't occur, including buf-

fer rows, bee-hive placement, and restricting distance from other orchards.

2. Because Arctic apples don't brown, they will disguise old/damaged fruit: The opposite is true! Arctic apples won't experience enzymatic browning (which occurs when even slightly damaged cells are exposed to air), but the decomposition that comes from fungi, bacteria and/or rotting will be just the same as conventional apples. This means that you will not see superficial damage, but you will see a change in appearance when the true quality is impacted.

Other accusations we hear somewhat frequently from a vocal minority who oppose all biotech foods are "we don't know what the effects will be down the road" or that we're "messing with God/Mother Nature". Regarding the first claim, the science tools we now have are truly amazing and we have an unprecedented level of precision, control and analysis when developing biotech crops. They must be meticulously reviewed before approval and around three trillion meals with biotech ingredients have now been consumed without incident. As to the messing with God/nature charges, biotech-enhanced crops are really just one more advancement in a long history of human-driven food improvements — and even the Amish and the Vatican support these advances!

Q5.

As an insider, you are privy to the goings-on and workings of the biotech industry, what do you envision the future of biotech to be? What new seeds are coming down the line and what potential advantages or disadvantages might they bring?

We foresee biotech continuing to be the most rapidly adopted crop technology ever, as it has been for the past 17 years. We also anticipate already realized benefits from biotech crops to continue, such as those highlighted by a fifteen year study including increased net earnings of $98.2 billion for farmers (mostly from developing nations), a reduction of 474 million kg of pesticide spraying and the equivalent reduction in greenhouse gas emissions as removing 10.2 million cars from the road for a year. Two major categories in particular where we'll see further advancements are in environmental sustainability (reduced pesticide use, carbon emissions, food waste) and higher crop yields under adverse conditions (from pest resistance, drought-tolerance, etc.).

Another major trend you'll see is the increased presence of biotech foods with direct consumer benefits, particularly nutrition. We will see many new projects following in the footsteps of crops like Golden Rice, which is fortified with beta-carotene; a precursor to Vitamin A. The World Health Organization has identified that around 250 million children under the age of 5 are affected by Vitamin A deficiency, which can cause blindness and death. Biotech crops like Golden rice can potentially save millions of lives by helping address this, and efforts are already underway to produce other Vitamin A enhanced crops including bananas and cassava.

This is just the tip of the iceberg, though, as there are many other exciting developments on the way including many other nutrient-enhancements for cassava, numerous drought-resistant crops, blight-resistant potatoes and many more. I actually highlighted some of these crops in a TEDx talk I gave in October 2012 on the value of agricultural biotechnology, which is available to watch online.

Q6.

As a biotech company, do you bear the brunt of the anti-GMO back-lash nominally directed at Monsanto and DuPont? If so, how has this affected you? Please be specific.

All companies who develop biotech crops have to deal with a certain level of backlash from the vocal, emotional minority who oppose biotechnology.

We are quite unique because when consumers discuss biotech companies, names like Monsanto and DuPont, as you mention, are the first ones that come to mind, rarely small companies like ours. Using Monsanto as an example, they have approximately 22,000 employees -- we have 7. Because most organizations in this industry are pretty massive, they do get the lion's share of attention. That being said, if we were to create a ratio of media attention to company size; ours would be through the roof!

One key reason we likely get more than our fair share of attention is that we're dealing with apples. When we're talking about something as popular and iconic as the apple (*e.g.*, "an apple a day", "American as apple pie"), it's going to get people emotionally charged. Genetically, our enhancement is relatively minor compared to the majority of crops out there; yet even so, when our petition was available for public comment along with 9 other biotech crops in the U.S., we received around three times as many comments as all 9 of the other petitions combined!

In terms of how all this attention affects us, we can dictate that to some extent. On one hand, we could simply choose to ignore it. The review process is evidence-based (and rightfully so!), meaning we could keep our heads down and let the science speak for itself and not worry about what people are saying. That's not how we operate, however, as we believe in the benefits and safety far too much to keep quiet. We want to do our best to make sure accurate, evidence-based information is out there to counter-balance all the myths and misinformation. This may mean that we spend more time and resources on education than others might, but it's too important of an issue not to.

We've made a concerted effort so transparency is the core of our identity. We know we have a safe, beneficial product and we're happy to explain the truth around previously mentioned misconceptions. We make it a priority, no matter how busy things get, to keep active on Twitter, Facebook, LinkedIn, maintain a weekly blog, make timely site updates, respond to every single sincere email we get and invest in delivering presentation such as last year's TEDx talk.

We believe everyone in the science and agricultural industries have a responsibility to help educate the public on the facts of biotechnology. Sometimes that results in more backlash, but it's worth it.

Q7.

Some scientists state that the anti-GMO backlash has cemented Monsanto's grip upon the market because only they can afford the regulatory burden, do you find this to be true in your experience? And how does this affect the greater biotechnology field?

Well, we've touched on how rigorous the review process is and how much smaller we are than the big industry players, so yes, it is tough for smaller companies to bring a biotech crop to market. It's challenging to raise funds, produce needed data, spend the resources providing education, and it's just a much bigger overall risk.

While the regulatory burden is heavier for small biotech companies, I think we're an example that it's still possible for the little guys to make it through, but it's not easy. Not only do you have to successfully develop a fantastic product, but you must be focused, persistent and very patient. There is no rushing the review process, but here we are a decade after first planting Arctic trees and we expect to achieve deregulation in the U.S. later this year.

Even though we're helping demonstrate it's possible for small companies to commercialize a biotech crop, the high regulatory burden certainly does affect the industry as a whole. With such an intimidating outlook in terms of high investment, both in time and resources, there will obviously be far less small, entrepreneurial companies than would be ideal. In a field in which innovation should be embraced as much as possible, we are missing out on many potential innovative companies and value-added products because the barriers are so high.

Really, what it comes down to is the regulatory process is (and should be) extremely rigorous, but it is indeed possible for companies that aren't multinationals to accomplish commercialization. Ideally, once biotech crops add further to their exemplary track record of safety and benefits and the scientific tools continue to improve; these barriers will gradually be lessened.

Q8.

Lastly, what is your relationship to the government and governmental agencies. It has been alleged that agencies like the FDA are in the pocket of big biotech organizations and are willing to look the other way. Do you find any truth in those statements? If not, why not?

If we had to select one word to describe the multiple regulatory bodies we've dealt with over the past few years (USDA, APHIS, FDA, CFIA) it would be "thorough". There's certainly no looking the other way and nothing casual about the review process. If these government agencies were in the pocket of biotech companies, we wouldn't still be awaiting deregulation more than ten years after we first developed Arctic apples!

Some people will see that some of the agencies have former members of biotech companies and immediately distrust the whole system; this misses the point. Of course they will have some former industry employees. These companies have thousands and thousands of employees and plenty of them are well-credentialed with first-hand experience in multiple facets of agriculture. In most fields, movement between private and public spheres is common, and most working aged citizens will have at least 10 different jobs before they turn 50. Some overlap is inevitable.

The truth is, you will hear a very wide range of arguments from those who don't like biotech crops and this is just another one on that list. Luckily, there is more than enough evidence to show that biotech crops are indeed safe and beneficial, including over 600 peer-reviewed studies, around one-third of which are independently funded. The best advice we can give to consumers is to do their own research, but always with a close eye on the credentials and reputability of the sources!

○§◎

NEAL'S BIO.

Neal Carter is president and founder of Okanagan Specialty Fruits™ (OSF), a biotechnology company specializing in the creation of novel tree fruit varieties. Outside of OSF, Neal and his wife Louisa grow and pack apples and cherries from their orchard in British Columbia's beautiful Okanagan Valley. For over 30 years Neal has worked around the globe as a bio-resource engineer, working on many different crops and projects ranging from maize to mango, from growing to harvesting, packing, storage, processing and packaging. It was through this firsthand experience that Neal was persuaded that biotechnology can help agriculture meet ever-expanding global food demand.

Neal founded OSF in 1996 in order to explore opportunities to utilize biotechnology to boost fruit consumption and growers' sustainability. OSF's flagship project is the development of non-browning Arctic® apples, which have been engineered to resist enzymatic browning by silencing the genes that produce polyphenol oxidase, the enzyme that drives the browning reaction. Arctic apples are currently progressing through the deregulation processes in both Canada and the United States and it is expected they will be available in grocery stores within a few years.

— Neal Carter

@ArcticApples

PART III.

Perspectives

Nothing in life is to be feared, it is only to be understood. Now is the time to understand more, so that we may fear less.

Marie Curie

WHY I GROW GMOS

BY JAKE LEGUEE

Farmer

What do you think about GMOs?

Is there a more emotionally charged question out there in this part of today's world? Certainly, it is understandable that the food we eat be an engaging issue for consumer and producer alike. There has been a drive from the consumer to learn about the food they eat. They want to know how it is produced, and whether it is in a sustainable fashion. Ultimately, and most importantly, they want to know if it is *safe*. An unknown factor like genetic modification is a cause for concern for these people, because the long-term effects are not readily available to us.

I support and applaud those in the public that ask these critical questions. Too many people don't think about the things that are done by the government, business and other organizations. The problem isn't in people asking questions; it is in people asking the *wrong* questions to the wrong *sources* – and believing the answers without question.

I am a farmer that grows genetically modified (GM) crops. Not all of my crops are GMOs. In fact, in a usual rotation of 5-7 different crops, only two are GMOs. Canola and soybeans, two of my farm's most economically important crops, are GMOs. Other crops, like wheat, peas and flax, are not GMOs, for there are simply none available. Contrary to popular belief, I *do* have a choice to buy GM crops or alternatives. So why do I grow GM crops when there are so many other cropping options?

That is a good question, and the answer will be different for every farm. In my life on the farm, canola and soybeans are our two newest crop options. In Western Canada, we have been growing wheat for as long as we've been farming. Flax and peas are old crops for us as well. Canola is one that we have really only been growing in earnest on our farm for the past 15 or so years. We only just started growing soybeans 3 years ago.

I suppose we could grow old open-pollinated canola and conventional soybeans (these are not GMOs). But would we do that? The claim I hear from some consumers is that GMOs are hazardous. By association then, I must be either cruel or naïve to grow these dangerous crops, putting other people at risk.

But here is the question I pose to the GMO haters: do you really believe I would grow these crops if I believed they were unsafe? My family eats the food we grow. I would not put them at risk if I truly believed GMOs were hazardous.

Honestly, I don't believe they are. GM crops are *not* danger-ous. In the almost 20 years since Monsanto started genetically modifying corn, soybeans and canola, the evidence has become clear that the benefits of genetic modification far outweigh the risks. This isn't an opinion by a biased industry representative. The information I use comes directly from **peer reviewed jour-nal articles**, the best source of information on anything scientific. GM crops also have dramatically reduced use of the most dangerous and volatile chemicals to control weeds. Most of the GM plants we deal with are "Roundup Ready", which means they are resistant to the active ingredient of Roundup, which is glyphosate. The way we measure the toxicity of chemicals like glyphosate is its LD50 number. This refers to the amount of the chemical, given all at once, which results in the death of 50% of the test animals. The acute Low Acute Toxicity for oral consumption of glyphosate in rats is an LD50 value greater than 5,000 mg/kg of body weight. This means that if you were a rat, and you weighed in at 3 kg, you would have to consume 15 grams of glyphosate for it to become toxic to you. That is quite a lot. Comparatively, the LD50 of caffeine is 192 mg/kg body weight. How much coffee do you have in a day? The point is, the dosage makes the poison, and any chemical can be toxic in a large enough dose.

Today's farm operation is a complicated business. Every year, we run through the numbers on each crop to decide which ones to grow and on how many acres. Canola and soybeans, and especially canola, are profitable crop options for us. So yes, we do grow these GM crops because they allow our farm to make money. Are they making us rich? I wish! But they do allow our farm business to make enough money to survive, and hopefully, over time, prosper. Is this not the dream for us all?

Ultimately, the question of why I grow GMOs comes down to the fundamental freedom that we all have in our democratic society: the freedom of *choice*. It is my choice to grow GM crops. Conversely, if you don't approve of them, it is your choice to buy something else. However, keep in mind the unintended consequences of doing so. GM crops allow us to use less toxic pesticides at lower rates. Furthermore, we can achieve unprecedented yields with the incredible biological advances made with these GM varieties. We need to grow 70% more food by 2050 to feed this growing world; we are going to need all the tools we can get to accomplish this.

My farm grows GM crops, and I am proud to say that we do. They are safe and sustainable crop options that we have the right to grow if we choose to. I hope that you will think about what I have said the next time someone asks you, "what do you think about GMOs?"

JAKES'S BIO.

Jake Leguee is a Co-Manager of Leguee Farms, a farm partnership that operates west of Fillmore. They grow wheat, canola, peas, soybeans, and various other crops. Jake also works as a sales agronomist with Top Notch Farm Supply in Fillmore, SK, and graduated with a B.S.A., specializing in Agronomy, from the University of Saskatchewan in 2010. As a farmer and an agronomist, agriculture, and the science and business therein, is his fascination and passion. This article in no way reflects the opinions or practices of Leguee Farms or Top Notch Farm Supply, Inc. and is my own.

— Jake Leguee

WE'RE ALL WEARING THE SAME GENES

BY ANASTASIA BODNAR

Plant Geneticist

Amajor "ick factor" of genetic engineering is that it allows us to take genes from one species and add them to another species. It's not something to be afraid of — in fact, as we learn more, it becomes more and more amazing. While it sounds strange, we are all wearing the same genes.

Look at the genome of any organism on the planet and you'll find at least some genes in common with other organisms. The root of this idea is evolution itself. We're all related!

You've likely heard that we are genetically 98.8% similar to chimpanzees. But did you know we are also 90% similar to cats, 80% similar to cows, 75% similar to mice, 70% similar to zebrafish, 60% similar to fruit flies and chickens, and 50% similar to bananas?

There are a few different ways that we can look at similarities between organisms, including phylogenetic trees, whole genome comparison, and individual gene comparison. All of these can help us to understand how similar we all are, and how genetic engineering is not icky at all.

Phylogenetic Trees.

Since we all have a common ancestor back in the primordial ooze, we all have genes in common. One way to look at this is in a phylogenetic tree. Like a tree, phylogeny starts with one trunk that then branches out into smaller and smaller branches. Organisms that are closer together on the tree diverged from a common ancestor more recently than organisms that are further apart on the tree. For example, snakes are more similar to lizards than they are to crocodiles.

While the similarities between organisms were originally determined by examining physical characteristics from bones to biochemistry, the advent of genome characterization and later sequencing has allowed us to better understand the similarities.

Whole Genome Comparison.

Another way to look at similarities is to compare whole genomes. The genomes of more than 200 organisms have been sequenced. The sequences can be aligned based on their similarities so we can see how similar the genomes are as wholes. During the evolutionary process, chunks of genome may be rearranged, mutated, duplicated, and changed in other ways, but we can still find the similar areas using software.

Individual Gene Comparison.

While we can see similarities at the whole genome level, looking at individual genes is useful, too. There are many genes in common across wildly different organisms. Some of them are conserved with amazingly few changes while others have mutated so much that we can just barely tell the genes had a common ancestor.

One example of this involves farnesene synthase, an enzyme that catalyses the synthesis of farnesene, which is a chemical compound that causes odor. Various forms of farnesene (and the enzyme that makes it) are found in many different organisms, including aphids and apples. In apples, farnesene contributes to that nice apple smell. In aphids, a slightly different farnesene is an alarm pheromone that tells other aphids to run away because a predator is near.

A version of the farnesene synthase gene was inserted into the wheat genome by Rothamsted Research in England with the goal of scaring aphids away. They also used a farnesyl pyrophosphate synthase gene, because farnesyl pyrophosphate is needed to make farnesene. Both genes were synthesized in a lab, with the goal of making genes that would work well in wheat and that would be effective at scaring aphids away. It just so happened that the synthesized farnesene synthase gene was most similar to the peppermint version of the gene. The synthesized farnesyl pyrophosphate synthase gene was most similar to the gene found in mammals, with one tiny sequence particularly similar to the cow version of the gene.

What all this means is that many genes appear in many different of organisms with minor differences. Homologous genes have a common ancestor, just as the organisms that the genes appear in have a common ancestor. It's not scary once you understand what's happening, and it's clear that the wheat hasn't been turned into mint or into a cow due to the addition of these genes.

Gene Movement Between Species.

Not only do we all have a lot of genes in common, there's also natural movement of genes between species. This is called horizontal gene transfer, in contrast to vertical gene transfer between an organism and its offspring.

One example is a new use of genes from a fungus. These genes allow the fungus to produce carotenoids, colorful molecules that make carrots and other plants yellow, orange, or red. Aphids, which are typically green, acquired the genes from the fungus so they can now appear yellow or red. These color changes affect which predators target the aphids (ladybugs like red aphids best while parasitic wasps like green ones), providing an evolutionary advantage to the aphids.

While examples of horizontal gene transfer like the aphids can be found, it seems to be a fairly rare phenomenon. Surely, as more and more genomes are sequenced, we will find more gene swapping. Still, there are restrictions. Bacteria swap genes fairly freely, and eukaryotes (multi-celled organisms like humans, aphids, plants, and fungi) can take up genes from bacteria and viruses (and very rarely from other eukaryotes). But many experiments have tested for gene transfer from eukaryotes to bacteria and found that it just doesn't happen. This means that genes put into plants through genetic engineering won't be taken up by bacteria. And, since some version of the genes exist in nature, bacteria already have access to the genes anyway.

What does this mean for GMOs?

Moving genes between organisms isn't any cause for concern. We can be sure that moving one or a few genes from one organism to another doesn't change the organism — a plant with a gene from a bacterium does not turn into a bacterium! Plus, the plant already had genes similar to ones from bacteria, and likely had genes transferred horizontally from bacteria, too.

One way to think about it is sentences in books. If we have a cookbook and a bible and we move a sentence from the cookbook to the bible — does it make the bible a cookbook?

There are already some sentences in the bible that are similar to those in the cookbook, and they have many words in common. Consider: "And you, take wheat and barley, beans and lentils, millet and emmer, and put them into a single vessel and make your bread from them." This verse is Ezekiel 4:9, the recipe behind Ezekiel Bread.

If we added a new cookbook sentence, it wouldn't change any of the other sentences. It would just add more information. It definitely would not transform the bible into a cookbook — just as adding a gene to wheat that is similar to a gene from cows does not turn that wheat into a cow.

It's not where you come from, but what you do with yourself.

One of the most common genetically engineered traits uses a gene from a bacterium called *Bacillus thuringiensis*. The gene is for a crystalline protein that is poisonous to certain types of insects, including some very important agricultural pests. The protein is called Cry toxin or Bt toxin. Whole Bt bacteria have been used in insecticidal sprays for many years, but the effect doesn't last long so it has to be reapplied often. It made sense to try expressing the gene for the Bt toxin in the plants you want to protect. The gene has been genetically engineered into many plants, including corn, cotton, and eggplant.

The Bt gene has allowed farmers all over the world to re-duce the amount of insecticide needed to protect their plants. It doesn't matter if the farm is big or small, insects can do a lot of damage. Many pesticides can cause harm, especially to non-target insects like butterflies and bees. Bt provides the best of both worlds — protecting food plants from insects and protecting the environment. Of course, nothing is perfect — any pesticide must be used carefully or pests will evolve resistance.

When evaluating genetic engineering, we have to look at what a gene does, not where it came from. The source of a gene doesn't matter — and many of our genes are the same anyway.

ᘓᘔ

ANASTASIA'S BIO

Dr. Anastasia Bodnar is Co-Director of Biology Fortified, a non-profit organization founded in 2008. Biology Fortified aims to provide science-based information about agriculture, including biotechnology and sustainability. Scientists like Anastasia work on Biofortified part-time while keeping up with the latest science in their day-to-day work. She has a PhD in genetics with a minor in sustainable agriculture from Iowa State University. Anastasia specialized in creating nutritionally enhanced crops through biotechnology and breeding. She previously worked in public health with the U.S. Army and in science policy at the National Institutes of Health. Anastasia now works as a Biotechnologist for the U.S. Department of Agriculture.

Disclaimer: The views and opinions expressed by Dr. Bodnar are solely her own and do not necessarily reflect those of the U.S. Department of Agriculture per 7 CFR 2635 subpart H.

— *Anastasia Bodnar*

@GeneticMaize

ON WHAT IT'S LIKE NOT TO KNOW SQUAT ABOUT GMOS

BY MIKE BENDZELA

Professor & Farmer

Our neighbors' beloved cow, Rosie, recently developed a prolapsed vagina; so they had her ground up, packaged up, and stored away in their freezer.

This is what farming entails.

Farming means learning lessons you didn't plan on learning, lessons you really didn't want to learn, and trying to turn these lessons into something good. This often involves killing, lots of killing.

And sometimes you have to kill not just mammals and arthropods but your own mislaid plans and bad ideas.

As we try to scrape food off of Maine's glacial till in a new farming venture I'm involved in, I'm continually shocked by my own ignorance. I will get slapped upside the head by some new, unforeseen difficulty and be forced to deal with it. I have to acknowledge that I'm part idiot, as Fourat says on Random Rationality.

For example, we just bought a sprayer from a neighbor whose husband, an old-time farmer, recently died. The machine didn't work, and I had to figure out why. I needed to take apart the diaphragm pump, not knowing a damn thing about diaphragm pumps. And so I did, and I bought a rebuild kit and cleaned all the crap out of the pump, packed it back together, and got the sprayer to dribble a little water. Now I'm in the midst of learning how to rebuild the "control assembly" so that the pump will actually build up pressure. What the hell, while I'm at it I might as well take apart the "pulsation damper" and see how that is, and of course the plastic dome inside is trashed, too. So I basically have had to learn how to rebuild the entire sprayer in a couple of weeks.

You learn as you go, and the sensation is often unpleasant — especially the sensation of realizing you don't know what the hell you're doing half the time. You frequently have to acknowledge that you are wrong, dead wrong, and that you can be absolutely stupefied in your wrongness — as I discovered in the last few years about the issues of organic farming and GMOs.

The problem with my assessing the facts on genetically-modified organisms is that the whole subject is way the hell over my head. So the issue becomes: Whom do I listen to? When I informally identified as an "organic" gardener and worked at an organic farm, I simply absorbed the prevailing orthodoxy without much thinking about it. The farm where I worked advertised "NO SYNTHETIC PESTICIDES! NO GMOS!" Surely there was a good reason for this. Surely they and the state organic certifying agency knew what they were talking about.

But two incidents conspired to shake me out of my stupor.

The first was engaging in an on-line debate years ago with Bob Carroll at the Skeptics Dictionary, whom I respect immensely, about his entry on "organic food/farming." The second was being trained as a pesticide applicator — to work for the organic farm.

I had a moment during that training where I literally said to myself, up inside my pea brain: "I'm at the state organic central headquarters being trained how to apply 'organic' pesticides lawfully and safely. If we have to follow the same rules as the 'conventional' guys, then what's the friggin' difference?"

Now I'm a licensed applicator, so I can speak a little to the issue of pesticides, how I've completely changed my mind about them (they're our friends). I will later try to extrapolate from my experience with pesticides to the issue of genetically modified organisms and explain how I came to change my mind about them as well.

I recently took over a class from a professor who could no longer teach it. Focused on issues related to food production and consumption, the course is designed to teach incoming freshmen the realities of academic life — like having to come to class and actually doing the work.

The university tapped me to take over the course because folks heard I "farmed" in the summer. I adopted some of the original readings and tried bending the focus of the course to what I knew from my experiences growing plants and animals over the years.

I concluded that many of these materials either trafficked in junk or were junk outright. The film "Forks Over Knives," a pseudoscientific paean to veganism, should be renamed "F*cked-Over Knaves". "Food, Inc.", a Gish gallop about everything that's "wrong" with the "Big Ag," should be called "Food, Ick!" I find them both egregious — silly, awful, just full of the most gosh-damnedest lies.

[The astonishing number of rave reviews this junk has received on Amazon is a testament to how willing we are to believe the most outrageous crap.]

I also got the opportunity to read Michael Pollan's "In Defense of Food," which I now refer to affectionately as "Indefensible Foodie." In this rant, he refutes the hated "reductionist science" straw man by appealing to . . . reductionist science. But anyway, this science can't explain why the god-awful "Western diet" (whatever that is) is killing us, so let's return to the Golden Era of grandma's "traditional" foods and the wisdom of holistic dentists.

I realized I was going to have to steer class discussions more toward critical thinking and debunking than toward the original theme of food production and consumption.

My new emphasis would not just be whether or not we should eat animals, for example, but how we as lay persons evaluate all sorts of claims about food and farming we see in the media. Is organic really to be preferred over "conventional" (whatever that is)? Do vegans lead better lives, and does eating meat really increase your risk of cancer? Is Big Ag trying to contaminate us with GMOs, poison us with pesticides, and infect us with *E. coli*? Are all our health problems — metabolic syndrome, Type 2 diabetes, obesity, high blood pressure, cardiovascular disease, autism, allergies — to be laid at the feet of the dreaded "Western diet"?

Yes, those are the various claims made in the materials I inherited for the class. I had to scramble during the semester to find ways to address the bull. I began with readings about critical thinking, starting with a chapter from Donald Prothero's textbook *Evolution* called "Science, Pseudoscience, and Baloney Detection." We also read a magnificent short piece called "Climate Science as Culture War," which discusses how our beliefs about scientific matters are dictated by our "ideological filters" and "referent groups."

I hoped to use these critical thinking tools to evaluate food claims. I scheduled a toxicologist to come in and give a talk about pesticide risks and benefits. We visited a friend's grass-fed beef operation to hear her explain why organic was not the way she wanted to go for the sake of the health of her animals. I hoped to convince the class not to believe everything they read and hear about food.

Nearly half the class either failed or dropped out. I felt I had worked too hard for nothing, but then I remembered those who did well and was told that this was pretty typical for freshman courses of this type.

While sitting recently through the tedium of hundreds of numbers being called while people were selected and sorted into panels for jury duty, I brought along a copy of an article on "gut microbiota" that a fellow old time musician and friend of mine wanted me to read. This friend and I share an interest in science writing, she being a bona fide scientist, me being a failed fiction writer-turned-adjunct instructor. I gathered she had liked the piece, but I found myself seething as I read it.

I recognized some familiar tropes — the flailing away at the "Western diet" straw man; the stringing together of wild-assed speculations with weasel phrases such as "could be the result of," "may help explain," "may be linked to"; the overall supercilious view of the products of "Western" science, all the while giddily citing scientific studies about the importance of gut bacteria. This guy is absolutely gaga over his new-found gut flora. He's even considering not washing his produce before he eats it. Then he says:

"Yet advising people not to thoroughly wash their produce is probably unwise in a world of pesticide residues."

Only someone wholly uninformed would write such a thing. Who is this guy? I turned back to the title page: Ah. I had been duped into reading Michael Pollan — duped because his name was in such tiny print on the cover page that I didn't even notice it.

Pesticide hysteria, I have discovered, is one of those ideologies that is perfectly immune to data. Anyone can find the pertinent data on the Internet. It's called the Pesticide Data Program, put out by the United States Department of Agriculture, and it's something I use in class to teach students that you simply cannot trust what you see in print media, on television, or on Internet sites about the alleged dangers of pesticides residues.

The PDP shows unambiguously that the nearly-universally loathed "Big Ag" farmers are doing a magnificent job keeping our food supply safe, that the "world of pesticide residues" exists only in the heads of urbanites who have probably never had to deal with pests eating away their livelihoods. The PDP documents that these dreaded "residues" come in infinitesimal quantities. They don't come anywhere near the tolerances set by the FDA.

So, for example, carbaryl, a very common insecticide that I use in my orchard, turns up in only 233 of 10,002 samples tested. The "worst" commodity turns out to be orange juice, with 130 out of 585 samples showing detections of carbaryl. But if you actually look at the figures you find that while the FDA tolerance is 10 parts per million, which is tiny, the range of detection in those samples of juice is between 0.003 and 0.018 ppm! In other words, between 3 and 18 parts *per billion,* or just barely above the level of detection of the sophisticated equipment used currently.

Think about that: while the law allows only minuscule tolerances of "pesticide residues" in foods like orange juice, with huge margins of safety built into those tolerances, the actual amounts detected are *orders of magnitude* below those tolerances. And this is true for all pesticides in all of the commodities tested by the USDA. This means farmers are well-trained; they're following label instructions, applying materials properly, and honoring "pre harvest intervals." And yet Michael Pollan thinks we live in "a world of pesticide residues."

Yet you should wash your produce. Why? *Food-borne illnesses!*

These are 100% organic, all-natural critters found on your food. Gut flora wannabes, if you will. It seems hard to believe, but the Centers for Disease Control and Prevention reports that in 2011 there were over *47 million* food-borne illnesses in the US, with 128,000 hospitalizations and over *three thousand deaths!*

For comparison's sake: The National Pesticide Information Center's latest annual report (June 1, 2011-May 31, 2012) says that of the approximately 3,000 human pesticide "exposure" incidents, fully three-quarters happened in "home or yard." Only 5% were "agriculturally related." 46.8% reported no symptoms. Nobody died.

If these are the figures for raw exposures during pesticide applications, then how many deaths and injuries do you think resulted from the 3 parts per billion of carbaryl "residues" found in orange juice? I'll bet you the answer's a big fat Zero.

So, yes, Michael, wash your damn organic greens, but knock off the apocalyptic rhetoric about "pesticide residues."

It's only a very well-fed country that tolerates seeing such mud flung at its farmers while dwelling on the bacteria in its gut.

Learning that pesticide hysteria was a load of bull primed me to learn that the anti-GMO movement, too, is just a group of howling hysterics jazzed up to look science-y.

It didn't take long: All it took was learning that the Humulin my Type 1 diabetic partner takes every day to keep himself alive is the product of recombinant DNA technology — a GMO. You isolate the sequence in the human genome that codes for the protein insulin, insert this DNA into the nucleus of an *E. coli* bacterium, and the bacterium divides to become a little insulin factory.

In my freshman class, I introduce the subject of genetic modification by typing "GMO images" into a Google search engine and marveling at the results on the screen. It's hilarious. I ask the students what images they notice the most, and they immediately see them:

"Hypodermic needles!"

There they are, sticking out of tomatoes like porcupine quills, being injected into ears of corn, filling apples with gory red fluids. Then I ask a simple question: "How is the hypodermic needle employed in the creation of genetically modified organisms?"

No one knows the answer.

"It isn't," I tell them.

Then we look at all the scary photo-shopped pictures: of frogs with orange peel for skin; of the tomato with a fetus in its pulp; of the heart-shaped potato dripping blood.

"How many of these are actual GMOs?" I ask the students. Of course, they recognize that these images aren't truthful. "What do you call that?"

"Call what?" a student asks.

"When you misrepresent something to make it look bad." "Propaganda?"

That's an acceptable answer. There is hope after all.

The next step:

"Does anyone know what a real GMO looks like?"

Someone mentions "soy beans." How would we find a real GMO? You can't rely on googling "GMO" to give you the answer. You have to know what you're looking for.

I type in "Humulin" and find an image of a boring little box. It's funny that it has a hypodermic needle near it, but I explain what it is and how it works.

I type "Rainbow papaya" into the search engine, and we look at the results on the screen — a gorgeous orange fruit engineered to resist ring spot virus, a disease that nearly decimated the papaya industry in Hawaii.

I type in "Golden Rice" and tell the story of how this rice has been altered with a gene sequence from a daffodil to express beta-carotene, the building block for Vitamin A, and how it could be used to help prevent blindness in impoverished children but how activists have held up its deployment.

I type in "Arctic apple" and explain that this Granny Smith apple has had a couple of genes tweaked to shut off enzymes that make the apple pulp turn brown. Not a big deal at first glance, unless you're an apple vendor who would like to sell freshly-cut apples as snack food.

These images are not anywhere near as arresting as the bleeding heart potato, and that's a shame. They're modest, simple pictures of food grown by real farmers.

We watch a video about the "spider goat" that produces a special silk protein in its milk. An orb-weaving spider's gene is inserted into the goat's genome, and this allows humans to "harvest" the silk protein from the goat and turn it into a durable material that can be used as sutures in surgery.

The Maine legislature is about to consider a bill that would require that all "GMOs" be labeled, so there is some talk about it in the news and amongst students' friends, most of whom are anti-GMO. Stupid letters and editorials appear in the newspapers.

"Truthfully, I don't care whether labels appear on genetically modified foods or not," I tell the students. "It might be interesting to see the law pass, though, because I think it will backfire. The labels will appear and most people just won't care. The activists clearly want the labels to deter people from buying these foods, but I don't think it will work. These foods are clearly not dangerous."

Yet some students have heard about the cancerous rat "studies," about that corporation, what's its name ... the one excoriated in "Food, Ick!"...

"Anyone can find a study for anything," I tell them. "The question is, have you read those studies yourself, or are you just parroting them? *Can* you read them? I confess, I've tried to read some of those studies and they go right over my head, and I have a Master's in English!

"Then you have to ask yourself whether you're equipped to *evaluate* those studies, and the answer is absolutely not! None of us can tell whether a study is a good study or a bad one, whether or not they used the proper controls, et cetera.

"We're lay people and what do we know? I'm not impressed by these studies. There are thousands of them, and you can go on the Internet and cherry-pick all day long until you find the ones that support your pre-existing beliefs.

"Remember what Dr. Harriet Hall says: 'Never believe one study. Always find out who disagrees and why.' If you haven't done that work, well then, shut up about it."

We're built to believe, not disbelieve, and it will be your life's work to decide what beliefs to accept and what to discard, and it won't be easy.

So 'be very, very careful,' as that Thomas Cardinal Wolsey quote says, 'about what you put into that head of yours, because you will never, ever get it out.'

"Beliefs are like fishhooks, easy to swallow, nearly impossible to cough up."

OK? Class dismissed.

ঙ৪

MIKE'S BIO

Mike Bendzela is an adjunct instructor of English at the University of Southern Maine, teaching introductory courses in composition, creative writing, and literature. He designs his courses to instill critical thinking skills in incoming freshmen, covering such topics as biblical criticism, the history of Darwinian thought, and contemporary writing about food and farming. He has a hidden history as a fiction writer, having published short stories in literary journals and winning a Pushcart Prize back in 1993. He gave up writing in 1999 to take up American old time fiddling and banjo, and gigs periodically with his band Spruce Rooster. For twelve years he was a volunteer Emergency Medical Technician in his town, taking a leave of absence to go into farming full-time. His domestic partner of 28 years and he co-own with their landlords a very small vegetable farm and orchard CSA (Community Supported Agriculture), growing a wide variety of crops for local subscribers. He learned that nothing dissuades a person from "organic" practices like trying to grow heritage apple varieties in dark, fungal Maine.

– *Mike Bendzela*

USING TECHNOLOGY TO REDUCE OUR FARMING FOOTPRINT

BY GABRIEL CARBALLEL

Farmer

Farmers believe that every day should be for the environment, because we depend on the environment to produce the food the world demands. We need good plants, good soil, and good weather. Without a good environment, we're helpless.

That's why we must take advantage of opportunities such as World Environment Day. Think of it as a second Earth Day. Each June 5, the United Nations sponsors WED. This year of 2013 theme was "Think. Eat. Save." Organizers have a specific request: "reduce your footprint."

Here on my farm in Uruguay, that's what we do all year round, thanks to advances in technology.

My family farms almost 6,000 hectares (roughly 15,000 acres) near the town of Mercedes, in Soriano. Our most important crops are soybeans, but we also grow corn, sorghum, wheat, barley, canola, oats, and grass seeds. The weather is variable but we never see snow, which allows us to plant for 12 months.

We started growing GM crops 16 years ago. It became obvious immediately that they're excellent for conservation.

As the website for WED points out, agriculture accounts for 80 percent of the world's deforestation. This is the result of pressure to convert wilderness into farmland, to keep pace with a booming global population. To protect what remains, we must produce more food on less land — and that's exactly what biotechnology lets us do.

The first year we planted GM soybeans on our farm, in 1997, we tried it on 30 hectares. The results were amazing. Within two years, we had converted entirely to GM soybeans. When biotechnology came to corn in 2004, we quickly switched to it as well. Genetic enhancement drove our yields upward because these excellent crops are so good at fighting weeds and pests.

Our experience shows that science can help us produce more with less — the very definition of sustainable agriculture.

There are other benefits as well. We're now able to do a much better job of maintaining natural pastures for a combined crop-cattle operation. This helps us preserve biodiversity.

Best of all, however, is our no-till farming system. Soil erosion is a huge challenge for farmers around the planet, but our soil is actually improving each year. Our crops pump carbon into the soil, and we can keep it there because we no longer need to fight weeds by tilling the soil after harvesting. At the end of the growing season, we simply leave the straw on top of the soil.

A friend of mine, Carlos Crovetto of Chile, puts it well: "Grains are for the people, straw and residues are for the soil."

GM crops make this possible.

Here's another statistic from the WED website: Agriculture is responsible for 30 percent of the world's greenhouse gas emissions.

And here's another benefit of biotechnology: Because we plant GM crops, our greenhouse gas emissions have dropped sharply. We're doing our part to combat climate change.

I can plant all of my fields with just two big tractors, an air drill, a planter, a big sprayer, and two combines. I've seen much smaller farms that use a lot more equipment, spewing out carbon emissions at a far higher rate than we do.

GM crops allow us to reduce the number of times we have to drive over our fields, which means that our environmental footprint has shrunk.

It's like we've reduced our shoe size. When does that ever happen?

Other advantages are harder to spot but they're equally real. Consider tire wear. I can buy a tractor, use it for 8,000 hours, and sell it with the same tires. This is important because petroleum is an important ingredient in tire manufacturing. The more use we can get out of our tires, the better — it's good for my bottom line as well as for the environment.

Unfortunately, many nations resist biotech crops because they don't understand the benefits. Farmers like me in South America already know why GM farming makes sense, as do farmers in the United States, Canada, and elsewhere.

If we're going to continue reducing footprints around the world, what we must do is spread the word.

CBEO

GABRIEL'S BIO

Gabriel Carballal farms with his father in Mercedes, Uruguay, growing soybeans, corn, wheat, barley, canola, oats, grass seeds, sorghum and raise beef. Gabriel is a member of the Truth About Trade & Technology Global Farmer Network.

Truth About Trade & Technology (TATT) is a non-profit advocacy group led by farmers who support freer trade and a farmers freedom to choose the tools, technologies and strategies they need to maximize productivity and profitability in a sustainable manner.

- Gabriel Carballal

@TruthAboutTrade

DO GMO CROPS HAVE A HIGHER YIELD?
IT DEPENDS

BY MICHAEL SIMPSON

Biochemist

The Union of Concerned Scientists (UCS) is an American environmental organization founded in 1969 at the Massachusetts Institute of Technology, which claims 400,000 members. They focus, generally, on environmental issues like nuclear power, global warming and a few other issues. Many of these issues are critically important, and a science advocacy group like UCS helps keep the scientific facts about global warming and other environmental issues at the forefront of the discussion.

But one area where UCS has gone off the rails of scientific evidence and embraces generally left wing science denialism is agriculture, more specifically GMO, or genetically modified organisms (or in this case crops). They are generally supportive of organic farming (which has little or no health benefit at a high cost to consumers) and vehemently opposed to GMO crops, based on what appears to be the same bad scientific critical skills that we observe in global warming deniers. There is nothing more frustrating than dogmatic science that stands against evidence.

There are few, convincing, peer-reviewed studies that show any risk from consuming GMO foods. Late in 2012, an article was published by Séralini *et al.* that seem to show that a particular GM corn, made by Monsanto, would cause cancer. The anti-GMO world was so desperate to grab onto any evidence that would support their beliefs about the evil of GMO crops that they broadcast this study widely. The dean of pseudoscience, Dr. Oz, hyped the study to his fawning and uncritical audience. Except the study was thoroughly debunked by a vast range of scientists, most of whom had little general interest in agriculture, but saw bad science for what it is: bad science.

A lot of the controversy about GMOs seems to be based on the precautionary principle, which states that if an action or policy has a suspected risk of causing harm to the public or to the environment, in the absence of scientific consensus that the action or policy is harmful, the burden of proof that it is *not* harmful falls on those taking an act. Yet, there are literally dozens of peer-reviewed articles that show that GMO crops are safe. And the scientific consensus also concludes that GMOs are safe. Once again, individuals conflate a political debate or opinion with scientific evidence. Other than a small group of scientists, most of them associated with UCS and other left-leaning environment groups, there just isn't a controversy with regards to GMO crops. Just switch sides of the political aisle, and it's the same thing with global warming or evolution — the only debate is a political one where right-wing science deniers are insisting that the vast mounds of scientific evidence are wrong.

It's also clear that anti-GMO feelings arise, partially, from the Appeal to Nature logical fallacy, that is, natural is better, even without any evidence supporting that belief (see my comments above about organic farming). What is amusing is that the natural, genetically unmodified corn, called teosintes, looks like your typical lawn grass. The fruiting body, the ear of corn that we all eat, is tiny. Corn was domesticated 10,000 years ago through, unsurprisingly, genetic modification. Our distant relatives were amazingly adept at genetic manipulation, and central Americans were able to domesticate corn. So the "natural" corn exists only as a wild plant in parts of Mexico. What we eat is completely different thanks to genetic modification, whether its grown in a huge agribusiness farm or in your small domestic cornfield in your backyard.

Moving beyond the fears of potential health danger of GM foods, what are the benefits? Probably the only reason to plant GMO crops is to vastly increase yields of food. This yield may be increased by reducing damage from pests, increasing drought resistance, or improving the amount of food from each plant. If there were no benefits from GM crops, then maybe the precautionary principle makes sense.

Of course, the UCS has a point of view on this matter. They published a white paper (a non-peer-reviewed document that tries to look like a real scientific paper) that concluded that "overall U.S. corn yields over the last several decades have annually averaged an increase of approximately one percent, which is considerably more than what Bt (a type of GMO corn) traits have provided." In other words, based on this one type of GMO corn, UCS is claiming that the corn doesn't have a higher yield than conventional corn (whatever that may be).

But the UCS is using a term called "intrinsic yield" which means something very specific in agriculture, and does not mean what many of us think is meant by "yield." The UCS claims that GMO plants (in this case corn) do not appear to produce higher "intrinsic yields"; but what they mean is that the

GMO crops don't produce more kernels per cob. But "yield" to a farmer means something more. It's the total amount of corn (or other crop) that they can produce at fixed costs (land, water, pesticides, whatever) in a fixed area. For example, if more corn survives to maturity because it is more resistant to pests and requires less pesticide (or other non-"natural" compounds), the yield of the farmer's field is higher, even if each individual plant does not produce more.

A recently published communication in *Nature Biotechnology* shows that GMO corn sometimes has higher and sometimes lower yields than conventionally bred corn, if you ignore all confounding factors in the environment. In years where GM corn was producing similar or lower yields than conventionally bred corn, the environmental factors, such as weather, disease or pests, were average. When accounting for the bad environmental situations, GM corn had significantly greater yields.

In other words, semantics matter. Using the UCS definition of "yield", which just looks at a single plant and ignores all other factors, GMO corn has no advantage. But in the real world of agriculture, the yield can be larger, sometimes quite a bit larger, under real world conditions that include a whole host of environmental challenges for the plant.

The point is that the value of GMO crops should not be underestimated, and the semantics can change how we value these crops. A real skeptic looks at the evidence for the value of the GMO crops (seems positive) while examining the evidence for the health risks (there is just nothing out there that scientifically supports any health issues with GM crops) — the scientific conclusion remains the same that GMO crops have a large positive benefit to mankind.

I know that a lot of hatred of GMO crops is pointed at Monsanto, which is one of the larger marketers of GM crops. But since many of the comments about Monsanto are strawman arguments or are intentionally poisoning the well, logical fallacies that are laughably similar to the arguments made about Big Pharma and vaccines, it's hard to accept them. There are some arguments about GM crops that have some validity. Biodiversity is one that is concerning, but that can be overcome with small, sustainable farms that are willing to produce genetically diverse crops, which will attract a higher price from consumers who want them. But in a crowded world with less and less fertile farmland, it is important that "yields" be increased, and that may always require genetic manipulation — something that was done 10,000 years ago to get us the first domesticated corn.

ℭ℘

MICHAEL'S BIO

Michael has over 25 years experience in marketing, business development, and product development in the medical products industry, working in a variety of marketing, sales, clinical research, and product development roles with large and small medical products companies. He has also had key executive roles on both the manufacturing and distribution sides of the medical products industry.

Michael has an undergraduate degree in Biology from a top US research university, and a graduate degree in Biochemistry/ Endocrinology from a major US research university, and did his post-graduate work in a multi-national pharmaceutical company.

— *Michael Simpson*

@SkepticalRaptor

WHY ORGANIC ADVOCATES SHOULD LOVE GMOS

BY RAMEZ NAAM

Author

What if there was a way to farm that spared the rainforests, cut down on toxins in our soil and waters, and provided healthier, more nutritious food?

Sounds like organic farming, right? But actually, it's GMOs.

The goals of organics — farms that cause less damage to the environment and grow food that's better for you — are great. But organic isn't living up to that potential.

In terms of nutrition, the consensus of multiple analysis of all the data (like the metastudy from Stanford and a 50 year systemic review from the UK) is that it's more or less a wash. Organic foods, in general, are neither more nor less nutritious than their conventional counterparts.

In terms of environmental impact, one might think that organic farms are the clear winners. And if you look at what happens on an acre of organic farm land vs. an acre of conventional farm land, that's correct. But an Oxford University meta-analysis of 71 peer-reviewed studies showed that, because organic farms use more land to grow the same amount of food, they erase their environmental benefit and are in some ways worse than conventional farming.

Save the Forests.

But even this is under-estimating the impact of organic farming, because the study above didn't look at the biggest issue of agriculture — the conversion of land from forest to farm. We use nearly 1/3 of the land area of the planet to grow food. That, in turn, has led to the destruction of half the original forest on the planet. Around the world, agriculture drives a whopping 80% of deforestation today. That destruction of forest is by far the worst environmental impact of agriculture, many times worse than the impact of pesticide or fertilizer over-use.

Meanwhile, projections are that by 2050, we're going to need to grow 70% more food around the world than we do today. If we did that by maintaining yields exactly as they are and spreading farms, we'd chop down 70% of the world's remaining forests. Trying to feed the world starting from organic yields would be far worse, because their yields are lower.

How much lower? In 2008, the USDA surveyed every organic farm in the US, asking about their yields. Plant pathologist Steve Savage compared those yield numbers to yield from conventional farms in the same years. Here's an excerpt from his summary:

In the vast majority of cases national organic average yields are moderately to substantially below those of the overall, national average.

Examples for row crops include Winter Wheat 60% of overall average, Corn 71%, Soybeans 66%, Spring Wheat 47% and Rice 59%.

A totally separate analysis, from researchers at the University of Minnesota, published in *Nature,* found that organic farms grow only around two-thirds of the same amount of food, per acre, as conventional farms, meaning that they need one and a half times the land of conventional crops.

The goals of organic are noble, but there's simply no way to feed the world with yields so low, unless we're willing to chop down all the forest that remains. Sparing forest means growing *more* food per acre, not less.

More Foods, More Forest.

How do we grow yield? We could do it by lifting worldwide yields up to US levels. That would mean giving farmers in the developing world better access to fertilizer, pesticides, and irrigation that drive yields up in the US. Of course, organic advocates would prefer *not* to use more fertilizer and more pesticides.

Is there another way? Perhaps — and GMOs may be key to that. So far GMOs have contributed only modestly to yield increases, but on the horizon are approaches that could make a big difference.

Consider the yields of corn (the most grown crop in the US) vs. those of rice and wheat (the two most important crops for food supplies globally). Corn grows about 70% more calories per acre than rice or what. Why? Because it has a newer form of photosynthesis called C4. Now, funded in part by the Gates Foundation, the C4 Rice Project is looking to port the genes for C4 photosynthesis to rice. Other projects are looking at doing the same for wheat. Those would essentially be rice and wheat varieties with a tiny bit of the corn genome in them (about 0.1%). And they could lift yields by more than 50% on their own, and more in combination with other advances. They would also reduce water and fertilizer needs of rice and wheat.

So — more food, less deforestation, less water need, and less need for synthetic fertilizer. Doesn't that align with the goals of organic advocates? And is it really profoundly unnatural to create strains of rice and wheat that borrow just a little bit of corn's genome?

Better for the Planet.

Organic advocates also want less pesticide use, in part to reduce toxicity to the environment. Ironically, GMOs are already doing this.

The National Academies of Science report *Impact of Genetically Engineered Crops on Farm Sustainability in the US* says this in the summary:

> When adopting GE herbicide-resistant (HR) crops, farmers mainly substituted the herbicide glyphosate for more toxic herbicides.

Glyphosate (roundup) has a nasty reputation, but in reality, it's dramatically less toxic than older pesticides like atrazine. And Roundup Ready crops have allowed glyphosate to almost completely replace atrazine on those fields. How much less toxic is Roundup than atrazine? About 200 times less toxic.

Other GMO work on the horizon could address another complaint organic farmers have about conventional farming — the heavy use of nitrogen fertilizer that runs off and creates dead zones. GMO farming has already reduced runoff by encouraging no-till farming. But a more radical project is underway. Legumes like peas and soy don't rely on nitrogen in the soil for fertilizer. Instead, with the help of friendly microbes, they extract nitrogen from the atmosphere, where it makes up 78% of the air we breathe. Another Gates-foundation funded project is looking at ways to give cereal crops — wheat, corn, and rice, for instance — that same ability to fertilize themselves from the air.

Aren't all of those things improvements?

Better for the People.

Finally, there's the health impact. Organic advocates want food that's more nutritious. And they're skeptical of the safety of GMOs. Yet the scientific consensus is that the GMOs we've approved for human consumption are entirely safe. Indeed, that consensus is at least as strong as the scientific consensus on climate change. Almost all GMO safety hysteria comes from a single media-manipulating lab, in France, which has had its work torn to shreds. Against that, hundreds of scientific papers have found GMOs safe. Looking at all that data, the American Association for the Advancement of Science concludes that GMOs are safe. So does the American Medical Association. So does the European Commission. Even the French Supreme Court threw out France's ban on a GMO because the French government couldn't produce any credible evidence that GMOs were a threat to the environment or human health.

More importantly, GMOs aren't just safe, they could boost nutrition. The Golden Rice project, which engineers rice to produce vitamin A in the edible grain (not just the leaf) could help 250 million children who have Vitamin A deficiency. (And for those fearful of corporate control over crops — Golden Rice will be free to virtually all farmers in the developing world, and freely-replantable. Every biotech company involved, including Monsanto, has waved their patent rights in the developing world.) Beyond Golden Rice there are many more enhanced nutrition projects in the works.

Inspired by golden rice, a team of Australian researchers in 2011 created an experimental rice breed that boosts vitamin A and also quadruples the amount of iron and doubles the amount of zinc in rice grains. An international team has taken the same ideas and applied them to Africa's most common staple crop, cassava, which feeds 700 million people, and created BioCassava, a variant that has increased levels of vitamin A, iron, and dietary protein.

So the next generation of GMOs could boost nutrition, reduce nitrogen fertilizer use, and boost yield, letting us feed the world without chopping down its remaining forest. Indeed, it's easy to imagine 'bio-organic' farms that don't use synthetic pesticides or fertilizer, but that do use these genetically enhanced seeds.

Environmentally cleaner, better for the forest, more nutritious, and able to feed the planet. Aren't those traits every organic advocate, every environmentalist, and, heck, every *person* in the world should welcome?

∞

RAMEZ'S BIO

Ramez Naam is a computer scientist who spent 13 years at Microsoft. He's also the award-winning author of three books. His latest, *The Infinite Resource: The Power of Ideas on a Finite Planet* charts a course to overcome the real challenges of climate change, feeding the planet, and a host of other natural resource and environmental threats.

- Ramez Naam
@RamezNaam

SCIENCE IS LAUGHING AT US

BY JULEE K

Teacher

As an former anti-GMO blogger, I knew in March that I had to come to a Moment of Truth after reading *The Lowdown on GMOs with Kevin Folta*. I was then forced to come clean about something that deeply troubled me. I read. I learned. I talked to real scientists. I even tried to get past the abstract in a report and decipher scientific jargon. I pondered. I soul searched. I changed. What conclusion did I come to?

Here it is: **Science is laughing at us**.

I put the statement in bold and used it as a title for this chapter because I want it to be seen, known and remembered. Science sees us as a joke. The "us" I am referring to are the folks behind the anti-GMO movement. I had been one of those folks for several months. The movement is gaining momentum in the US right now and I had been solidly in the camp of banner waving followers, but my views are starting to moderate because of what I learned and continue to learn.

Let me be clear about one thing. By science, I do not mean the biotech industry or *any* industry. I do not mean Monsanto, DuPont, Dow or any of those huge multinational corporations. I mean pure *science* for the purpose of science. The kind that is studied and practiced at institutions of higher learning where people get doctorates and do cutting edge research. Of course, industry may *apply* the research but what I'm getting at is that the science I'm talking about does not have a monetary or social agenda. It is not biased. After all, if it were biased, it could not be science.

There is a great divide going on, between science and — let's call it *non-science*. Oh sure, there has always been contention in this arena but today, this contention affects whole populations and the livelihood of individuals. It affects food, land and water resources. It affects politics, policies and very important decisions that cover both. Scientists are laughing at us — but they're also crying because movements like this wield a lot of power and well meaning folks can, unbeknownst, do more harm than good.

Now I'll get specific and to the point. Scientific evidence, the real kind, the kind that is peer-reviewed and published in respectable journals does NOT show harm from eating genetically modified food. Let me repeat: Scientific evidence does not show harm from eating genetically modified food. Science tells us that the benefits outweigh the risks.

I can't put it any simpler than that. I am not saying that there aren't problems with the technology. There are many, as with *any* technology, and no one says it is perfect in its present state but the bottom line is, the genetic engineering of crops and eating food from those crops is not, in and of itself, unsafe, and we can push against it all we want but I promise you this — it is not going anywhere. Just ask anyone at any time that tried to stop progress!

With that anti-GMO chapter of my life closed, I began a new way of looking at the biotech industry. I am willing now to try to understand the scientific side of genetic engineering. I will learn as I go.

Does all this mean that I'm going to rush to the store and buy boxes of processed, GMO laden food? Certainly not! I will still eat organic, whole foods as often as possible. It is my choice to do that. And no doubt, my visceral response to GMOs will take a lot longer to change than my intellectual one. But I'm growing, learning and changing and that is always a good thing. Being *willing* to challenge my beliefs was the pivotal moment.

Have I gone over to the dark side? Well, I'm doing this because I have to if I want to look at myself in the mirror. I want to be a beacon of truth. I cannot, in good conscience, continue to spread propaganda.

The tipping point for my "conversion" came from the recent "Stunning Corn Comparison" that went viral worldwide. The report was said to be independently conducted by a major food company at the request of a grower. Originally circulated by the organization *Moms Across America*, the report claimed shocking differences between the nutritional value of conventional and GM corn and even claimed that GM corn was toxic. I remember first seeing the report and, truth be known, questioned its validity, but in my eagerness to be a good soldier for the movement, I immediately published it on my blog.

Well, guess what happened? I got spanked by a scientist and you know, I deserved it. The report is bogus, and apparently, this would be apparent to any undergrad who passed organic chemistry 101. This wasn't the only piece of information circulated by the anti-GMO movement that got caught in my logic crosshairs but it was certainly the most glaringly ridiculous. It was being repeatedly circulated on every activist website I could find. And people were believing it. It got to the point where every time I saw it I wanted to barf.

Bottom line: This is not a movement that I can take pride in belonging to anymore. I want to base my opinions on facts, not fallacies. I named my blog *Sleuth4Health*. Here is one definition for sleuth: *to carry out a search or investigation in the manner of a detective.* Well, that is what I am doing, will continue to do, and I don't know where it will lead but I must follow the evidence, as any good inspector would do.

On the other hand, I'm still in support of labeling efforts — mandatory labeling, voluntary labeling, any kind of labeling (beyond what is currently available, which isn't much). I do feel the cat is out of the bag on this issue and we can't stuff it back in.

What I'd really like to see happen, however, is *real* education for the masses on this topic so we may all know what is true and what is not and make the appropriate choices for ourselves and for our families. People might be pleasantly surprised to learn, for example, that the GM corn or soy product they are eating actually made it possible for less pesticide to be used, not more, *because* it was genetically engineered.

And meanwhile, I am still *very much committed* to writing about and exposing real toxins — whether in our food, environment, or elsewhere. I just don't want to bark up the wrong tree.

೦೩ಙ

JULEE'S BIO

Julee K launched the blog *Sleuth4Health* in September of 2012 with a completely anti-GMO mindset. Outraged at what she perceived to be unwelcome guests at her table, GMOs, she was out to educate the world about the dangers of biotechnology and Sleuth4Health was born. With a BA and some graduate work in music and the humanities behind her, Julie's science background consisted of a combined four terms of geology and astronomy. Still, as she placed herself in the center of the anti-GMO movement and its rhetoric, something didn't quite ring true for her and faithful to the name of her blog, she began sleuthing.

What she discovered after several months of digging came as a complete surprise even to her. She has been enthusiastically blogging about it ever since. Ever the musician and music educator by profession, she hopes that by researching and posting regularly on her blog, she can continue to expand her knowledge in the sciences.

- Julee K
@sleuth4health

A PRINCIPLED CASE AGAINST MANDA-
TORY GMO LABELS

BY MARC BRAZEAU
Writer

I am opposed to government mandated GMO labels, though I started off in favor of them. In fact I helped a little on the campaign for labeling when I was working for Hartford Food System in Connecticut. Once I developed a stronger understanding of the issues surrounding genetically engineered crops, I realized that, not only do mandatory GMO labels make no sense, but they go against my principles.

Many people have a hard time wrapping their heads around how anyone could be opposition to a government mandated label for foods with ingredients derived from crops bred using the techniques of genetic engineering. They tend to assume that there is no principled case to be made and that all the opponents of mandatory GMO labels must have some financial stake in the issue. (I do not. In fact, not being opposed to genetically engineered crops narrows my horizons as a progressive writing about the food system.)

Critics of GE crops will ask, "Well then, what is wrong with asking for a simple label. How is that too much to ask? After all, don't we have a right to know what's in our food? How can you possibly be against labeling GMOs?"

There actually is a principled, common sense case to be made against mandatory GMO labels, but there are a few things we need to get out of the way before getting to that.

First of all, about that 'right':

"People are usually surprised to learn that there is no legal right to know," said Michael Rodemeyer, an expert on biotechnology policy at the University of Virginia in Charlottesville. A variety of rules and regulations control the words that appear on food packages. Such rules must be balanced against companies' constitutionally protected right of commercial speech, experts said. "It's an unsettled area in the law," said Hank Greely, director of the Stanford Center for Law and the Biosciences in Palo Alto. "If I were a betting man, I think the odds are good that the Supreme Court would … strike down a GMO labeling requirement."

I would argue that people do have a "right to know" what is in their food, but that government isn't always the proper vehicle for mediating that right.

To understand why someone would oppose a mandatory label identifying GE ingredients, you first have to understand the philosophical case against government overreach when it comes to commercial speech. This is laid out quite clearly in the four part test established by the Supreme Court in Central Hudson Gas & Electric Corp. v. Public Service Commission of New York.

Since 1980, the courts have analyzed regulations affecting advertising for commercial products or professional services (i.e., commercial speech) under the four-part test set forth by the U.S. Supreme Court in Central Hudson Gas & Electric Corp. v. Public Service Commission of New York. The "Central Hudson" test asks:

1. whether the speech at issue concerns lawful activity and is not misleading;

2. whether the asserted government interest is substantial; and, if so,

3. whether the regulation directly advances the governmental interest asserted; and

4. whether it is not more extensive than is necessary to serve that interest.

In this analysis, the government bears the burden of identifying a substantial interest and justifying the challenged restriction: "The government is not required to employ the least restrictive means conceivable, but it must demonstrate narrow tailoring of the challenged regulation to the asserted interest — a fit that is not necessarily perfect but reasonable; that represents not necessarily the single best disposition but one whose scope is in proportion to the interest served."

Or as the court found in International Dairy Foods vs. Amestoy:

Accordingly, we hold that **consumer curiosity alone is not a strong enough state interest** to sustain the compulsion of even an accurate, factual statement, … (compelled disclosure of "fact" is no more acceptable than compelled disclosure of opinion), in a commercial context.

A mandatory GMO label is likely to fail the test of Central Hudson because it does not address anything misleading nor does it address the state's interest in communicating relevant health, safety or nutrition information. GE crops have been shown to be substantially and compositionally equivalent to their conventional counterparts. There is no credible* evidence that GE crops aren't nutritionally equivalent to their conventional counterparts and pose no greater risks. Because the certified organic and non-GMO labels already exist, it would be difficult to show that a mandatory label is not more extensive than what is necessary to serve any substantial interest, if one was shown. Without any credible scientific evidence for a substantial government interest, we are left with mere 'consumer curiosity' and the courts have told us that isn't enough.

All of that tells why a mandatory label is likely to be found unconstitutional by a court (especially the Roberts court). It doesn't tell us why we might oppose such a law.

Here are five reasons why I oppose mandatory GMO labels.

1. This is not the proper role of government I personally think that Central Hudson provides a common sense framework not just for judging whether the state HAS the right to compel speech, but whether it SHOULD. Because there are an infinite number of things that consumers COULD be curious about, we need some criteria for judging when the government SHOULD step in to address that curiosity.

A mandatory label simply is not a common sense or principled use of state authority. For someone who cares about seeing policy that is grounded in principle and effective in accomplishing its goals, a mandatory GMO label is a disaster. A government-mandated food label should provide the consumer with relevant, actionable health, safety, and nutritional information and/or it should help the consumer from being misled. I have to agree with the courts that mere consumer curiosity is not sufficient to justify government involvement in product labeling. The burden is on advocates to explain why mere "consumer curiosity" should provide the basis for government mandated labeling. I haven't heard that issue being addressed yet.

As I said, the things that a vocal majority of citizens might be curious about are endless and we need firmer principles as to when to invoke the power of the state.

The only two rationales that I've seen put forward by proponents are "It's popular" and "Transparency". Using the popularity of something to justify public policy is hardly worth addressing. It should be self evident that the fact that something is popular is not a firm foundation for establishing public policy. A variation on this theme is that "64 other countries have labeling, why can't we?" Plenty of bad policies can be found in a collection of 64 or more countries. Again, popularity is not a justification for public policy. If 64 other countries jumped off a bridge…

As to "Transparency," a mandatory GMO label does nothing to create transparency. Being transparent with information requires that the information is relevant and substantial. A mandatory GMO label tells us nothing about which ingredient(s), which trait(s), it tells us nothing about the properties of the ingredients and how they differ from non-GE ingredients. It tells us nothing about other breeding techniques that someone may be curious about. People say they are concerned about pesticides, and yet non-GE crops often require more pesticides and the label would tell us nothing about pesticide use either. It certainly doesn't tell us if any of the ingredients are necessarily from herbicide-resistant crops, since there are also non-GE herbicide resistant crops.

What's more, these laws have huge carve-outs that render the transparency argument laughable. Most GE crops produced for food are fed to livestock, and yet meat is exempt. Most Americans consume a huge proportion of their meals in restaurants and yet restaurants are exempt. There are other nonsensical exemptions, but those are the big ones.

Nor does a mandatory GMO label provide information that consumers my have about other breeding techniques. Mutagenic breeding, where plant tissue cultures are exposed to radiation or harsh chemicals to induce (hopefully) useful traits is far more prone to unintended consequences than the precise methods of genetic engineering. Yet, mutagenic breeding is exempt from labeling. Clone grafting in fruits leads to monocultures in orchard fruits, bananas and grapes. Because these crops are genetically identical (far less genetic diversity than we see in corn and soy) more pesticides are needed in their cultivation than say, corn and soy. Yet, clone graft breeding is exempt from labeling. Run of the mill selective breeding has been responsible for taking common food crops and making them dangerous for humans. There have been a number of mishaps, but the most notorious was the Lenape Potato in the 1960s which ended up toxic as the alkaloid solanine was dialed up beyond tolerable levels in the quest for a better potato chip. Someone with concerns about mutagenic, clone graft breeding, or even selective breeding gone awry is not served by a mandatory GMO label. Singling out one breeding technique from the others doesn't provide transparency, it obscures other potential risks. This is not to say that any of these techniques are particularly risky, just to say that they all carry similar, very small risks. There are risks associated with all plant breeding. The risks associated with breeding using the techniques of genetic engineering are as small as those of other techniques.

Furthermore, ingredients like oils and sugars have no proteins and no genetic material. Versions of these products are chemically identical to those derived from crops bred by other methods. What does a label tell you in this case? Nothing about the ingredients or properties of the product. "But," you may protest, "it's not the DNA I want to avoid, but being implicated in the use of excessive herbicides." I hate to break it to you, but when a conventional farmer moves from RoundUp Ready canola to a non-GE alternative, it's likely to be Clearfield sunflowers or canola which are bred to withstand applications of imazethapyr. Are you sure you prefer imazethapyr to glyphosate?

The bottom line is that if people have a right to know things about their food that don't relate to health, safety, nutrition, or the prevention of fraud, then a voluntary label is the correct vehicle to mediate that right. People have a right to know if their food is kosher or halal, but in those cases, government is not the correct vehicle to mediate that right. Instead, that right is mediated for the consumer by private third-party certification in a voluntary system. The same principles apply for addressing consumer curiosity about genetically engineered ingredients.

2. Government has enough on its plate As a progressive, I want government to do a number of things really well. What I don't want is for government to try to do every single thing that the citizenry can dream up and then do it poorly. In Oregon, the state struggled to make the website for the Oregon Health Plan work. I really want my governor and his administration focused on doing what they have already been tasked with well. Not seeing how many plates they can get spinning at once. The certified organic label and the Non-GMO label are already providing the relevant information for those wishing to avoid GMOs and they already have the necessary infrastructure to carry out their mission. The state of Oregon does not currently have that infrastructure. I don't see any reason to further stretch an already overextended government. I don't imagine I need to provide a much different explanation for conservatives.

3. A mandatory GMO label is potentially misleading
Since government has traditionally mandated only useful and important information, a mandatory GMO label could potentially mislead some (not all) consumers into thinking that a GMO label contains useful and important information. Government usually doesn't single out ingredients unless there is a health or safety concern, like trans fats (before they were removed) or wheat and peanuts for those with allergies. Many may construe a front of the package label as a warning label. And in fact, the other countries, the ones label proponents are so fond of touting, don't have front of the package labels. The labeling is integrated into the ingredient label. Telling that this isn't the model being proposed, no? Since "64 other countries" already do it that way. As was pointed out above, singling out a single set of breeding techniques implies that other breeding techniques carry no risks. That is dishonest.

Simply, I don't believe the government should mandate misleading labels.

4. A patchwork of state legislation will be a nightmare
Has everyone forgotten the Articles of Confederation? A patchwork system of state laws governing commerce didn't turn out very well the first time we tried it. Is there any reason to believe that approach is going to work any better today?

How is a single state supposed to inspect and certify product with complex supply chains entering not only from other states, but all over the world? Where does the money for enforcement come from? How are companies supposed to try to comply with inconsistent and contradictory labeling regimes in different states? This seems like a completely unworkable, poorly thought out Rube Goldberg approach to a (non) issue. It's hard enough to make well-conceived public policy work. Starting from this basket case is a recipe for failure.

5. The environment Finally, the reason most animating to me, personally, is that a mandatory GMO label would likely lead to an increase in the use of non-GE crops in conventional agriculture. This would lead to a large increase in the use of soil-applied insecticides and an increase in the environmental impact of herbicides used. It would also lead to a decrease in no-till cultivation. That would result in a loss of carbon sequestration, increased erosion, and decreased soil fertility. That would be a very bad thing for the environment.

I just want food labels based on science and a firm understanding of the proper role of government. Is that too much to ask?

♋

MARC'S BIO

Marc Brazeau is an essayist and editor for the website *Food and Farm Discussion Lab,* as well as the founder and administrator of the online community of the same name. Combining interests in science, economics, politics, labor, agriculture, cooking, and nutrition with previous careers as a chef, restaurant owner, food security activist and union organizer, he provides a unique perspective on the food system issues.

One of his central concerns is in how a lay person makes sense of contested issues in a complex world. Understanding the ways that biases and motivated reasoning warp our discussions and and understanding of the food system comes front and center in his work.

He lives and works in Portland, Oregon.

- *Marc Brazeau*

@RealFoodOrg

GOOD, KINDHEARTED PARENTS ARE PRO-GMO

BY KAVIN SENAPATHY

Writer

As a parent, my worst nightmare is the thought of harm coming to my children. I cannot fathom the heartache of one of them suffering a debilitating illness. More unimaginably terrifying would be seeing the death of one of my children. As President Obama said after the Sandy Hook school shooting, "Someone once described the joy and anxiety of parenthood as the equivalent of having your heart outside of your body all the time, walking around. With their very first cry, this most precious, vital part of ourselves — our child — is suddenly exposed to the world."

Whenever I enter this rabbit-hole of catastrophizing, I pause and thank the universe for my family's privileged existence. My family is healthy. We have a beautiful home. We have everything we need and a lot of what we want. We can afford to send our daughter to an excellent preschool. I have the financial and professional flexibility to work from home a few days a week and revel firsthand in my 1-year-old son's developmental milestones.

I can only imagine how it feels to live on the edge of financial disaster, or to worry about my child's physical well-being. Like most kindhearted and empathetic people, my heart breaks for those less fortunate. Like many self-proclaimed liberals and democrats, I'm pro-welfare, pro-social programs, and pro-affordable and government-subsidized healthcare.

This is why I simply cannot comprehend why so many liberals, selfless in so many ways, are anti-GMO. Yes, yes, I know. Corporations. But that is not a valid argument when it comes to this:

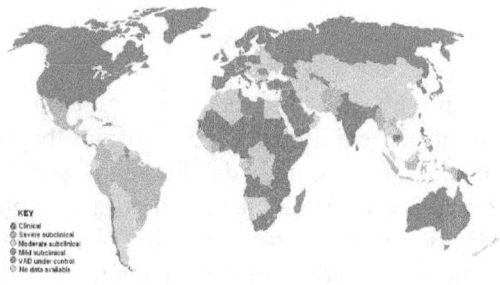

Image credit Wikipedia

This is a world map of the prevalence of Vitamin A deficiency based on WHO data. See that Americans? We in our cushy existence don't have to deal with this. While we're here marching against Monsanto, pretentiously shopping at Whole Foods, and berating Chobani for using GMO feed, many in those red and orange nations on the map are living off of less than what a serving of Chobani costs.

They are also watching their precious children — as precious to them as ours are to us — suffer from dreadful micronutrient deficiencies like VAD. Vitamin A deficiency is the leading cause of blindness in children, and greatly inhibits immune system function. Furthermore, VAD is a leading cause of maternal mortality worldwide.

So why is there opposition to GM rice and bananas with the potential to eradicate this horrific scourge? If you're unfamiliar with these crops, Golden Rice and more recently 'super bananas' have been modified to synthesize and accumulate beta carotene, which is converted to vitamin A when metabolized by the human body. Because of the higher concentration of beta carotene (which gives modern carrots their orange hue), the color becomes yellow or orange. Populations that could benefit consume these crops as staples. For example, Indians consume large amounts of rice, and Ugandans large amounts of banana. While nutritional deficiencies cannot be eradicated without improving infrastructure and social/political issues, these engineered crops have huge potential for tangible outcomes.

So why does Greenpeace vandalize Golden Rice test fields? Why pull the race card with disingenuous drivel about the super banana being a ploy to force GMOs into a brown peoples' market? As a 'brown person,' I can honestly say that I've looked suffering in the malnourished eye on my regular visits to India. The nerve of using the race card to oppose these potentially life-saving GM crops is detestable. I've said time and again, GM technology is safe. I urge everyone to read about the basic science and benefits of genetically engineered foods. In the meantime ask yourself, if anti-GM proponents truly thought GMOs were unsafe, why care only about elite brands like Chobani or stores like Whole Foods? Don't they care about the proles who can only afford Yoplait?

In the same speech I mentioned earlier, Barack Obama said, "This is our first task — caring for our children. It's our first job.

If we don't get that right, we don't get anything right. That's how, as a society, we will be judged." Our society, my friends, is the whole world. Good people are their brothers' keepers, the keepers of their sisters' children. Let's look past the blinders of privilege, reach above a lack of scientific awareness, and drop the empty anti-corporate ideology and see the light.

ೞ

KAVIN'S BIO

Kavin Senapathy was born in Washington D.C. and lives in Madison, Wisconsin. Kavin works for a genomics/bioinformatics R&D and contributes to <u>Grounded Parents</u> (a Skepchick site), <u>This Week in Pseudoscience</u> and other sites. She loves all things genetics, genomics, and bioinformatics. Her interests span the human and agricultural realms. Opinions expressed are her own and do not reflect her employer.

- *Kavin Senapathy*

@ksenapthy

GORILLAS AND THE FUTURE OF CROP BIOTECHNOLOGY

BY STEVE SAVAGE

Plant Pathologist

There are some really cool improvements coming along in several crops that have been developed using the tools of biotechnology — GMOs if you will. Some of these innovations have consumer health benefits. Some expand ways to encourage greater produce consumption. Some reduce food waste. Some prevent crop losses through disease and reduce the need for copper sprays. These traits represent an expansion of biotech beyond the major row crops primarily grown for animal feed or for fiber to crops like apples, oranges, tomatoes, pineapples and potatoes. Whether these new options actually make it to consumers depends a great deal on decisions that will be made by gorillas. I don't mean the kind of gorillas that Jane Goodall studied. I mean the kind in the expression "eight hundred pound gorilla."

The Eight Hundred Pound Gorillas of the Food Industry.

In many industries, there are players with disproportionate economic leverage who are often referred to as the eight hundred pound gorillas for their sector. In the food/beverage industry, there are huge entities from importers (e.g. Dole, Chiquita), to manufacturers (e.g. Mars or Frito-Lay), to food service retailers (e.g. McDonalds or Starbucks), to grocery retailers (e.g. Safeway, Wal-Mart...) who have an out-sized influence on not only their market segment, but on their supply chain. In the early years of crop genetic engineering, entities like these used their influence to slow or stop the commercialization of biotech crops. The question is what role these gorillas will play for the next generation of potential crop improvements.

When the Gorillas Embraced Biotech.

When genetically engineered crops were first being commercialized in the mid-1990s, several of the gorillas were supportive. Frito-lay was funding its own potato genetic engineering effort focused on storage life and chip quality. Dole and Chiquita were both in discussions with biotech companies about potential solutions to their biggest disease challenge (Black Sigatoka disease) and about the potential to make a banana with longer counter-life at optimal ripeness. Forward-thinking folks at Starbucks were looking into whether they might need to get involved in helping their small coffee bean producers through horticultural research and extension. Genetic engineering was one of the topics on the table. Meanwhile, farmers in the 1990s who grew crops with biotech options were very happy with what these options offered them (soybeans, corn, sweet corn, cotton, canola, squash, potatoes, papayas).

When Gorillas Are Weak.

However, by the later 1990s and early 2000s, the efforts of anti-GMO groups to demonize the technology were beginning to take effect. They used dramatically misleading imagery (great big hypodermic needles full of mysterious colored liquids, fruit and vegetables with faces...) and anti-corporate conspiracy theory rhetoric to alarm consumers. For all their protestation, there was no change in the scientific assessment of the intrinsic safety of the technology, nor were there real cases of health or environmental problems. However, the manipulated trend in consumer thinking began to worry the gorillas. The large players in the food system have tremendous power and influence, but they also have great vulnerability to anything that could tarnish their consumer brand. Activists of all stripes have taken advantage of this for many causes, and the anti-GMO forces began to do the same. Most of the continuing GMO crops are sold into non-branded animal feed and food manufacturing ingredient channels, so they were little effected by brand protectionism. For crops that do flow to consumer-branded companies, the story was different.

Round One Didn't Go So Well.

Starbucks was one of the first companies to come out with a pledge not to use GM (even though there were no GMO options for coffee anywhere close to the market). Unfortunately they also dropped the whole supply line horticultural support idea as well. The big banana companies backed away from any GMO projects. The major candy companies used their influence to delay for many years the introduction of herbicide tolerant sugarbeets. Even though the corn hybrids grown for the chip market were not generally GMO, the marketing side of Frito-lay reactively pledged that they would not use GMO corn for their chips.

The company then quietly dismantled their GMO potato effort. At one point, McDonald's made three phone calls to their major frozen fry suppliers asking if they could get only non-GMO potatoes. That effectively ended the growing of biotech potatoes - an improvement which required far less insecticide use. In the conspiratorial narrative of the anti-GMO movement, Monsanto is portrayed as being so powerful that it is able to "control the food supply." In reality, neither Monsanto or the entire potato growing and processing industry could do anything once the 800 pound gorilla of spuds decided it didn't want to risk brand-tainting protests. The gorillas were subdued by the activists in the first round of biotech innovation.

What About Round Two?

So, it has been at least 10 years since that first round of gorilla influence. Why should anything be different with this next round of technology offerings? It is not as if the anti-GMO movement has become any less intent on its mission or any more honest about the science. Brand sensitive companies are still basically risk-averse. The general public is probably no better informed about the actual science or about the ever stronger scientific consensus supporting the safety and utility of these technologies. Some factors remain the same, but not everything.

What Has Changed.

News Sourcing: people don't get their "news" the way they did in the 90s. Most have chosen the sources that fit their worldview, so something like an anti-GMO protest is going to be covered, or not covered differently for segments of the public. Indeed for those who get their information from websites like Grist or Mother Jones, there is some sort of anti-food industry or anti-GMO diatribe almost every day. It is hard to imagine how some new development will stand out against that background. For the rest of society that hasn't drifted into full-blown conspiracy theory thinking about biotechnology there may be some fatigue from the fear-mongering. There must eventually be a statute-of-limitations for saying that the sky is falling.

Communication by Scientists: overall, scientists have not communicated that well with the broader public, but that has been changing to some degree. There are a number of websites that do a much better job than what was around in the late 90s. Meanwhile, the collected body of independent, peer reviewed publications supporting the safety of biotech has grown.

Communication from the Farming Community: the Ag community has increasingly begun to use the internet and social media to tell its story. They are understandably tired of being demonized and falsely represented by the «food movement.» This response includes everything from farmer bloggers to farming organizations.

Real Food Supply Issues: In the 90s it certainly seemed that food was in abundant supply — even in over-abundance. Since then there have been some shocks to the global food trade balance that are hard to ignore and there is some reason to believe that we are in a new paradigm between population growth, expanding Asian middle classes, and climate change. This has meant a bit of food budget stress for the rich world, but it has already had a major political role in events like the "Arab Spring."

Calling People Out About The Science: A number of voices have pointed out the fact that neither the political Right or Left has been consistent in respecting the scientific consensus. Those who embrace the consensus on climate change tend not to on biotechnology and visa versa. Mark Lynas, a past anti-GMO campaigner, articulates this extremely well as have some in the mainstream press and in Books such as *Science Left Behind*.

A California Surprise: The voters of California soundly defeated a deeply flawed GMO labeling initiative in the 2012 election. This was completely unexpected since the "just label it" message originally sounded logical to 90% of them. The GMO labeling forces blamed this loss entirely on the level of spending by the food and biotech industry, but like Sheldon Adelson, they might need to acknowledge that votes are not simply for sale and that voters might actually have some independent, critical thinking abilities once they get the facts.

Can We Support Courage On the Part of the Gorillas?

Last year, Seminis Seeds (a Monsanto subsidiary) commercialized some new insect resistant sweet corn hybrids. Even though Syngenta's Bt sweet corn had already been on the market for many years, the major grocery retailers and processors had quietly suppressed its use and it was mainly grown for the roadside market. Thus few of the mostly local sweet corn growers ever got to take advantage of that technology which could have saved them many insecticide sprays each season. The anti-GMO crowd tried to make a big issue of the new hybrids and threatened to sponsor a boycott of Wal-Mart if they carried the product. Wal-Mart (an eight hundred pound gorilla if there ever was one) was bold enough to say that they saw no reason not to carry Bt sweet corn. Whether they actually did isn't clear. Still, the controversy faded.

Perhaps scientists, farmers and reasonable people in general can encourage the gorillas to take a different stand this time. For instance, I'd like to be first in line to buy "Arctic Apple" from some brave retailer and then pass them out to friends and family. With social networking organizing something like that with lots of supporters is certainly possible. Maybe things could start small with deliveries to a few distribution points in people's garages. How about coming together to enjoy some fries from whatever restaurant is first willing to talk about using a healthier oil to cook low acrylamide potatoes? How about writing campaigns to encourage gorilla companies to stand up to the purveyors of fear.

CR8O

STEVE'S BIO

Steve Savage is an agricultural scientist involved in agricultural technology for 32+ years. Originally trained as a plant pathologist, his career has taken him into many other disciplines and touched on many different crops and geographies. He likes to garden and has a little, 25 vine vineyard from which he makes wine each year.

Steve is passionate about meeting the challenge of feeding 9-10 billion people without destroying the environment. He believe that technology is a big part of how we will do that and is deeply concerned about the increasingly anti-science environment in which we live today. He doesn't think that scientists have done a good job of communicating and is doing his little part to address that problem.

- Steve Savage

@grapedoc

GMO OPPONENTS ARE THE CLIMATE SKEPTICS OF THE LEFT

BY KEITH KLOOR

Journalist

I used to think that nothing rivaled the misinformation spewed by climate change skeptics and spinmeisters.

Then I started paying attention to how anti-GMO campaigners have distorted the science on genetically modified foods. You might be surprised at how successful they've been and who has helped them pull it off.

I've found that fears are stoked by prominent environmental groups (such as Greenpeace), supposed food-safety watchdogs (like the Center for Food Safety), and influential food columnists (Mark Bittman among others); that dodgy science is laundered by well-respected scholars and propaganda is treated credulously by legendary journalists (Bill Moyers and Marion Nestle); and that progressive media outlets, which often decry the scurrilous rhetoric that warps the climate debate, serve up a comparable agitprop when it comes to GMOs.

In short, I've learned that the emotionally charged, politicized discourse on GMOs is mired in the kind of fever swamps that have polluted climate science beyond recognition.

An audacious example of scientific distortion came late in 2012, in the form of a controversial (but peer reviewed!) study that generated worldwide headlines. A French research team purportedly found that GMO corn fed to rats caused them to develop giant tumors and die prematurely.

Within 24 hours, the study's credibility was shredded by scores of scientists. The consensus judgment was swift and damning: The study was riddled with errors — serious, blatantly obvious flaws that should have been caught by peer reviewers. Many critics pointed out that the researchers chose a strain of rodents extremely prone to tumors. Other key aspects of the study, such as its sample size and statistical analysis, have also been highly criticized. One University of Florida scientist suggests the study was "designed to frighten" the public.

That's no stretch of the imagination, considering the history of the lead author, Gilles-Eric Seralini, who, as *NPR* reports, "has been campaigning against GM crops since 1997," and whose research methods have been "questioned before," according to the *New York Times*.

The circumstances surrounding Seralini's GMO rat-tumor study range from bizarre (as a French magazine breathlessly reports, it was conducted in clandestine conditions) to dubious (funding was provided by an anti-biotechnology organization whose scientific board Seralini heads).

Another big red flag: Seralini and his co-authors manipulated some members of the media to prevent outside scrutiny of their study. (The strategy appears to have worked like a charm in Europe.) Some reporters allowed themselves to be stenographers by signing nondisclosure agreements stipulating they not solicit independent expert opinion before the paper was released. That has riled up science journalists such as Carl Zimmer, who wrote on his *Discover* magazine blog: "This is a rancid, corrupt way to report about science. It speaks badly for the scientists involved, but we journalists have to grant that it speaks badly to our profession, too... If someone hands you confidentiality agreements to sign, so that you will have no choice but to produce a one-sided article, WALK AWAY. Otherwise, you are being played."

Speaking of being played, have I mentioned yet that Seralini›s book on GMOs, *All Guinea Pigs!* was published (in French) the same week? Oh, and there's also a documentary based on his book that came out simultaneously. You can get details on both at the website of the anti-biotech organization that sponsored his study. The site features gross-out pictures of those GMO corn-fed rats with ping-pong-ball-size tumors.

It's all very convenient, isn't it?

None of this seems to bother Tom Philpott, the popular food blogger for *Mother Jones*, who writes that Seralini's results "shine a harsh light on the ag-biotech industry's mantra that GMOs have indisputably proven safe to eat."

Philpott often trumpets the ecological and public-health dangers posed by genetically modified crops. But such concerns about GMOs, which are regularly echoed at other left-leaning media outlets, have little merit. As Pamela Ronald, a UC-Davis plant geneticist, pointed out last year in *Scientific American*: "There is broad scientific consensus that genetically engineered crops currently on the market are safe to eat. After 14 years of cultivation and a cumulative total of 2 billion acres planted, no adverse health or environmental effects have resulted from commercialization of genetically engineered crops.»

So what explains the lingering suspicions that some people (even those who aren't Monsanto-hating, organic-food-only eaters) still harbor? Some of these folks are worried about new genes being introduced into plant and animal species. But humans have been selectively breeding plants and animals pretty much since we moved out of caves, manipulating their genes all the while. The process was just slower before biotechnology came along.

Still, being uneasy about a powerful, new technology doesn't make you a wild-eyed paranoid. The precautionary principle is a worthy one to live by. But people should know that GMOs are tightly regulated (some scientists say in an overly burdensome manner).

Many environmentalists are concerned that genetically modified animals such as "Franken-salmon" could get loose in the wild and out-compete their non-engineered cousins, or lead to breeding problems for the wild members of the species. But even the scientist on whose research the "Trojan gene" hypothesis is based says the risk to wild salmon is "low" and that his work has been misrepresented by GMO opponents.

Another big concern that has been widely reported is the "rapid growth of tenacious super weeds" that now defy Monsanto's trademark Roundup herbicide. That has led farmers to spray their fields with an increasing amount of the chemical weed-killer. Additionally, some research suggests that other pests are evolving a resistance to GMO crops. But these problems are not unique to genetic engineering. The history of agriculture is one of a never-ending battle between humans and pests.

On balance, the positives of GM crops seem to vastly outweigh the negatives. A recent 20-year study published in *Nature* found that GM crops helped a beneficial insect ecosystem to thrive and migrate into surrounding fields.

The bottom line for people worried about GMO ingredients in their food is that there is no credible scientific evidence that GMOs pose a health risk.

Even Philpott, in his charitable take on the Seralini study, admits that, "no one has ever dropped dead from drinking, say, a Coke sweetened with high-fructose syrup from GMO corn." In the next breath, though, he wonders: "But what about 'chronic' effects, ones that come on gradually and can't be easily tied to any one thing? Here we are eating in the dark." Despite the study being a train wreck, Philpott's takeaway is that it "provides a disturbing hint that all might not be right with our food — and shows beyond a doubt that further study is needed." What's beyond a doubt here is Philpott's unwillingness to call bullsh*t when it's staring him in the face.

I single out Philpott not to pick on him, but because he represents the most reasonable, level-headed voice of the anti-GMO brigade (whose most extreme adherents don white hazmat suits and destroy research plots). The same goes for *Grist*, which calls the French study "important" and says "It's worth paying attention to what Seralini has done."

Such acceptance by lefties of what everyone else in the reality-based science community derides as patently bad science is "just plain depressing," writes a medical researcher who blogs under the name Orac. He compares the misuse of science and scare tactics by GMO opponents to the behavior of the anti-vaccine movement.

The anti-GM bias also reveals a glaring intellectual inconsistency of the eco-concerned media. When it comes to climate science, for example, *Grist* and *Mother Jones* are quick to call out the denialism of pundits and politicians. But when it comes to the science of genetic engineering, writers at these same outlets are quick to seize on pseudoscientific claims, based on the flimsiest of evidence of cancer-causing, endocrine-disrupting, ecosystem-killing GMOs.

This brand of fear-mongering is what I›ve come to expect from environmental groups, anti-GMO activists, and their most shamelessly exploitive soul travelers. This is what agenda-driven ideologues do. The Seralini study was seized on by supporters of California›s Proposition 37, a voter initiative, that wasn't successful in November of 2012, to require most foods containing genetically modified ingredients to be labeled as such in the state.

What›s disconcerting is when big media outlets and influential thought leaders legitimize pseudoscience and perpetuate some of the most outrageous tabloid myths, which have been given fresh currency by a slanted 2011 documentary (Bitter Seeds) that is taken at face value at places like the Huffington Post.

In a recent commentary for *Nature*, Yale University's Dan Kahan lamented the "polluted science communication environment" that has deeply polarized the climate debate. He writes: "People acquire their scientific knowledge by consulting others who share their values and whom they therefore trust and understand." This means that lefties in the media and prominent scholars and food advocates who truly care about the planet are information brokers. So they have a choice to make: On the GMO issue, they can be scrupulous in their analysis of facts and risks, or they can continue to pollute the science communication environment.

ೞ

KEITH'S BIO

Keith Kloor is a freelance journalist and adjunct professor of journalism at New York University. His work has appeared in Slate, Science, Discover, Nature Climate Change, Archaeology, and Audubon Magazine, among other outlets.

From 2000 to 2008, he was a senior editor at Audubon Magazine. In 2008-2009, he was a Fellow at the University of Colorado's Center for Environmental Journalism, in Boulder, Colorado. At his Discover magazine blog, he covers a wide range of topics, from conservation biology and biotechnology to climate change and archaeology.

— Keith Kloor
@KeithKloor

TIME TO CALL OUT THE ANTI-GMO CONSPIRACY THEORY

BY MARK LYNAS

Historian

I think the controversy over GMOs represents one of the greatest science communications failures of the past half-century. Millions, possibly billions, of people have come to believe what is essentially a conspiracy theory, generating fear and misunderstanding about a whole class of technologies on an unprecedentedly global scale.

This matters enormously because these technologies — in particular the various uses of molecular biology to enhance plant breeding potential — are clearly some of our most important tools for addressing food security and future environmental change.

I am a historian, and history surely offers us, from witch trials to eugenics, numerous examples of how when public misunderstanding and superstition becomes widespread on an issue, irrational policymaking is the inevitable consequence, and great damage is done to peoples' lives as a result.

This is what has happened with the GMOs food scare in Europe, Africa and many other parts of the world. Allowing anti-GMO activists to dictate policymaking on biotechnology is like putting homeopaths in charge of the health service, or asking anti-vaccine campaigners to take the lead in eradicating polio.

I believe the time has now come for everyone with a commitment to the primacy of the scientific method and evidence-based policy-making to decisively reject the anti-GMO conspiracy theory and to work together to begin to undo the damage that it has caused over the last decade and a half.

On a personal note, let me explain why I am standing here saying this. Believe me, I would much prefer to live a quieter life. However, following my apology for my former anti-GMO activism at my Oxford speech in January, I have been subject to a co-ordinated campaign of intimidation and hate, mostly via the internet.

Even when I was at school I didn't give in to bullies, and at the ripe old age of 40 I am even less inclined to do so now. Moreover, I have been encouraged by emails and other support from globally-renowned scientists who are experts on this issue, and who all said basically the same thing to me: 'You think you've got hate-mail? Welcome to my world.'

I think these scientists are the unsung heroes of this saga. They carried on with their important work and tried year after year to fight against the rising tide of misinformation, while people like me were belittling and undermining them at every turn. I won't mention names, but they know who they are. Some of them are here today, and I would like to give them my deepest thanks.

So for me also there is also a moral dimension to this. The fact that I helped promote unfounded scare stories in the early stages of the anti-GMO movement in the mid-1990s is the reason why I now feel compelled to speak out against them. I have a personal responsibility to help put these myths to rest because I was so complicit in initially promoting them.

My activism, which I wrongly thought of at the time as being 'environmental,' has done real damage in the world. For me, apologizing was therefore only the beginning. I am now convinced that many people have died unnecessarily because of mistakes that we in the environmental movement collectively made in promoting anti-GMO fear. With that on your conscience, saying sorry and then moving on is not enough. Some restitution is in order.

Following a decade and a half of scientific and field research, I think we can now say with very high confidence that the key tenets of the anti-GMO case were not just wrong in points of fact but in large parts the precise opposite of the truth.

This is why I use the term conspiracy theory. Populist ideas about conspiracies do not arise spontaneously in a political and historic vacuum. They result when powerful ideological narratives collide with major world events, rare occasions where even a tiny number of dedicated activists can create a lasting change in public consciousness.

In the 1960s the conspiracy theories about Kennedy's assassination reflected the profound feeling that there were shadowy people high up in the CIA and government who were subverting democracy, and fighting the Cold War by devious and deadly means. More recently, conspiracy theories about 9-11 reflected the hatred many on the political Left had for the Bush Administration.

Successful conspiracy theories can do real damage. In Nigeria an outbreak of Muslim conspiracy theorising against the polio vaccination campaign there led to a renewed polio outbreak which then spread to 20 other countries just when the disease was on the brink of being entirely eradicated.

In South Africa during the presidency of Thabo Mbeki the HIV/AIDS denialist myth became official government policy, just as the anti-GMO denialist myth is official European Union policy today. The result in South Africa was that hundreds of thousands of people were denied life-saving anti-retroviral treatments and died unnecessarily.

The anti-GMO campaign has also undoubtedly led to unnecessary deaths. The best documented example, which is laid out in detail by Robert Paarlberg in his book 'Starved for Science', is the refusal of the Zambian government to allow its starving population to eat imported GMO corn during a severe famine in 2002.

Thousands died because the President of Zambia believed the lies of western environmental groups that genetically modified corn provided by the World Food Programme was somehow poisonous. I have yet to hear an apology from any of the responsible Western groups for their role in this humanitarian atrocity.

Friends of the Earth was one of those responsible, and I note that not only has no apology been forthcoming, but Friends of the Earth Europe is still actively promoting GMO denialism in the EU in a new campaign called Stop the Crop. Check out their YouTube video to see how they have learned nothing in ten years.

Another well-known example is that of Golden Rice, genetically modified to contain high levels of beta carotene in order to compensate for the vitamin A deficiency which kills hundreds of thousands of children around the world and blinds many more every year. One study on the prospects for Golden Rice in India found that the burden of vitamin A deficiency could be reduced by 60%, saving 1.4 million healthy life years.

Here the actions of Greenpeace in forestalling the use of golden rice to address micronutrient deficiencies in children makes them the moral and indeed practical equivalent of the Nigerian mullahs who preached against the polio vaccine – because they were stopping a lifesaving technology solely to flatter their own fanaticism.

I think this campaign is shameful and has brought the entire environmental movement into disrepute, with damaging consequences for the very beneficial work that many environmentalists do. Greenpeace's campaign against vitamin A-enhanced Golden Rice should therefore be cancelled, and I call on everyone concerned about children's health to lobby Greenpeace and demand that this happens immediately and without delay.

The anti-GMO campaign does not even have the benefit of intellectual coherence. If you truly think that herbicide-tolerant biotech crops are an evil plot by Monsanto to achieve a stranglehold on the entire world's food supply, why would you also oppose all other non-patented and open-source applications of biotechnology, which have nothing to do with Monsanto, apparently without exception? This is like being against all computer software because you object to the dominant position of Microsoft Office.

On a logical basis only a case by case assessment makes sense for deciding how any technology might best be applied. So if you think that Bt corn is bad for US farmers, despite all the evidence to the contrary, it shouldn't necessarily follow that you also have to ban virus-resistant papaya, or oppose a blight-resistant potato in Ireland.

This matters today more than ever because we are entering an age of increasingly threatening ecological scarcity. The planet is beginning to move outside the envelope of stable temperatures that we have enjoyed for 10,000 years, and into an age of instability and rapid change.

Within just a year from now, global CO_2 concentrations will break through the crucial 400 parts per million boundary, marking a change is atmospheric chemistry that is unprecedented for at least 3 million years.

Moreover, we are now on a global emissions path which puts us on track for 4-5 degrees Celsius of warming by 2100, a transformation which will leave this planet barely recognizable and considerably more hostile to human and other life.

But what about all those who say that global warming is a hoax, a product of thousands of scientists conspiring with governments and the UN to falsify temperature data and usher in a new age of global socialism?

Well, I've spent more than a decade arguing with climate skeptics, and in the end I fall back on a single killer argument: that if an overwhelming majority of experts say something is true, then any sensible non-expert should assume that they are probably right.

To make the point, here is the consensus position of the American Association for the Advancement of Sciences on climate change:

"The scientific evidence is clear: global climate change caused by human activities is occurring now, and it is a growing threat to society. Accumulating data from across the globe reveal a wide array of effects: rapidly melting glaciers, destabilization of major ice sheets, increases in extreme weather, rising sea level, shifts in species ranges, and more. The pace of change and the evidence of harm have increased markedly over the last five years. The time to control greenhouse gas emissions is now."

Oh, but wait — the AAAS has also released another statement of consensus science on another area concerning us today:

"The science is quite clear: crop improvement by the modern molecular techniques of biotechnology is safe... The World Health Organization, the American Medical Association, the U.S. National Academy of Sciences, the British Royal Society, and every other respected organization that has examined the evidence has come to the same conclusion: consuming foods containing ingredients derived from GM crops is no riskier than consuming the same foods containing ingredients from crop plants modified by conventional plant improvement techniques."

So, my suggestion today is that a sensible baseline position for environmentalists and indeed everyone else is to accept the consensus science in both these areas. Instead, you have the un-edifying spectacle of so-called green groups like the Union of Concerned Scientists stoutly defending consensus science in the area of climate change, while just as determinedly undermining it in the area of biotechnology.

Tellingly, the UCS utilizes the exact same techniques as climate skeptics in its enduring and strikingly unscientific campaign against GMOs: it issues impressive reports based on strategic cherry-picking and only referencing its ideological allies in a kind of epistemological closed-loop, it pushes the perspective of a tiny minority of hand-picked pseudo-experts, and it tries to capture and control the public policy agenda to enforce its long-held prejudices.

Many of the most influential denialists like those at the Union of Concerned Scientists sound like experts; indeed they may even be experts. Richard Dawkins tells a story about a professor of geology, who lectured and published papers about stratigraphy in hundred-million year old rocks whilst at the same time being a 'young-earth' creationist who really believed the world was only 6,000 years old. His pre-existing religious conviction simply overpowered his scientific evidence-based training.

An even more striking example is Peter Duesberg, the leading light in the AIDS denialist movement, who is a professor of cell biology at the University of California in Berkeley.

Many anti-vaccine campaigners, like Andrew Wakefield, started out as qualified medical professionals. This is why scientific consensus matters — it is the last line of defense we have against the impressive credentials and sciency-sounding language of those who are really on the lunatic fringe.

Speaking of the lunatic fringe, someone else who claims scientific credentials is Vandana Shiva, probably the most prominent Indian anti-biotechnology activist, who incidentally draws much larger audiences than this one to her fiery speeches about the evils of Monsanto and all things new in agriculture. Shiva tweeted after my Oxford speech that me saying that farmers should be free to use GMO crops was like giving rapists the freedom to rape.

That is obscene and offensive, but actually is not the half of it. Let me give you my all-time favorite Vandana Shiva quote, regarding the so-called terminator technology, on which she launches constant blistering attacks without once acknowledging the salient fact that it was never actually developed.

"The danger that the terminator may spread to surrounding food crops or the natural environment is a serious one. The gradual spread of sterility in seeding plants would result in a global catastrophe that could eventually wipe out higher life forms, including humans, from the planet".

Now, I've said and done some pretty stupid things in my time, but this one takes some beating. You don't need the intelligence of a Richard Dawkins or indeed a Charles Darwin to understand that sterility is not a great selective advantage when it comes to reproduction, hence the regular observed failure of sterile couples to breed large numbers of children.

As Shiva's case so clearly shows, if we reject data-driven empiricism and evidence as the basis for identifying and solving problems, we have nothing left but vacuous ideology and self-referential myth-making. Indeed in many related areas, like nuclear power, the environmental movement has already done great harm to the planet, even as it has rightly helped raise awareness in other areas such as deforestation, pollution and biodiversity loss.

Science tells us today that the coming age of ecological scarcity extends much further than just global warming. If we wish to preserve a semblance of current biodiversity on this planet, for example, we must urgently curtail agricultural land conversion in rainforest and other sensitive areas.

This is why organic agriculture is an ecological dead-end: it is dramatically less efficient in terms of land use, so likely leads to higher rates of biodiversity loss overall. Maybe organic producers should be legally mandated to specify on labels the overall land-use efficiency of their products. I'm all in favor of food labeling by the way when it comes to something important that the consumer should have the right to know.

Of course conventional agriculture has well-documented and major environmental failings, not least of which is the massive use of agricultural fertilizers which is destroying river and ocean biology around the world. But the flip side of this is that intensive agriculture's extremely efficient use of land is conversely of great ecological benefit.

For example, if we had tried to produce all of today's yield using the technologies of 1960 – largely organically in other words – we would have had to cultivate an additional 3 billion hectares, the area of two South Americas.

We cannot afford the luxury of romanticized but inefficient agricultural systems like organic because the planet is already maxed out in terms of both land and water. Our only option therefore is to learn to do more with less. This is known as sustainable intensification – it's about improving the efficiency of our most ecologically scarce resources.

But remember, everything is changing. Food demand will inevitably skyrocket this half-century because of the twin pressures of population growth and economic development. We need to sustainably increase food production by at least 100% by 2050 to feed a larger and increasingly affluent global population.

This is where the eco-Malthusians tend to pop up, illustrating another uncomfortable aspect of the anti-GMO philosophy. Let me share with you a rather revealing quote I read just a couple of weeks ago on Yale 360, from the US environmental writer Paul Greenberg, where he is lamenting the supposed wrongs of genetically engineered salmon. But forget the fish – when it comes to humans he says the following:

> "If we continue to bend the rules of nature so that we can provide more and more food for an open-ended expansion of humans on the planet, something eventually will have to

give. Would you like to live in a world of 15 billion people? 20 billion? I would not. And while it's possible you will label my response as New Age-ish, I feel that GE food distracts us from the real question of the carrying capacity of the planet."

Well, I think that calling these sentiments New Age-ish is to give them far too much credit. I would actually call them misanthropic. What Greenberg seems to be suggesting here, as Paul Ehrlich did before him, is the denial of food to hungry people in order to prevent them breeding more children and contributing to overpopulation.

Luckily this modern-day Malthusianism is wrong in point of fact as well as by moral implication. Firstly, the human population is never going to reach 20 billion. Instead, it is forecast to peak at 9-10 billion and then slowly decline.

Secondly, although we are certainly heading for 9 billion people by mid-century, but that is not because people in poor countries are still having too many babies. The main reason is that children who are born today are much more likely to survive, and become parents themselves.

It is a little-known fact that the global average fertility rate is now down to about 2.4, not far above natural replacement of 2.1. So pretty much all the increased population growth to 2050 will come from more children surviving into adulthood.

And that is surely a good thing. I want to see child death rates in developing countries continue to plummet thanks to better healthcare, access to clean water and sanitation, and all the other benefits the modern world can and should bring to everyone.

No doubt like all of you, I also want to see an end to the scourge of hunger which today affects more people in an absolute sense than ever before in history. It is surely an abomination that in 2013 we can all go to bed each night knowing that 900 million other people are hungry.

This scourge affects children disproportionately — one third of child deaths are attributable to malnutrition. Among those who survive, nutrient deficiencies like iron, zinc and vitamin A can lead to cognitive impairment and other health problems, reducing a child's life chances for his or her entire future.

It is a truism to say that people are hungry not because there is a global shortage of food in an absolute sense, but because they are too poor to afford to eat. But it is a dangerous fallacy to suggest therefore that because the world on average has enough food, we should therefore oppose efforts to improve agricultural productivity in food insecure countries.

In fact probably the best way to address rural poverty is to ensure that subsistence farmers the world over enjoy more reliable and increasingly productive harvests. This will enable them both to feed their own families and to generate a surplus to sell at a profit so their children can go to school.

Is genetic modification a silver bullet way to achieve this? Of course not. It cannot build better roads or chase away corrupt officials. But surely seeds which deliver higher levels of nutrition, which protect the resulting plant against pests without the need for expensive chemical inputs, and which have greater yield resilience in drought years are least worth a try?

And real-world evidence so far gives grounds for optimism. The use of Bt cotton in China has been shown to dramatically improve biodiversity, unlike broad-spectrum insecticides which kill everything, pests and predators alike. The Bt protein only affects the insects which bore into the crop, is entirely safe for us, and has led to insecticide reductions of 60% in China and 40% in India on cotton.

The introduction of Bt brinjal in India, a project which I know people here in Cornell were closely involved in leading, would have dramatically reduced insecticide poisonings associated with that crop too, had the anti-GMO activists in India not succeeded in preventing its use.

India today seems to be perched on a scientific knife-edge, with a vociferous lobby pushing dark-age traditionalism on the brink of permanently capturing the entire political and legal agenda. If they succeed, hundreds of millions of food-insecure Indians will be the losers.

In Africa too there are a multitude of western-funded NGOs who all claim to be mysteriously protecting biodiversity by keeping cultivated plant genetic improvements permanently out of the continent. In many African countries GMOs are subject to the same kind of de-facto ban as is the case in Europe, leaving poorer farmers at the mercy of a changing climate and exhausted soils.

However, a showdown is looming, because some of the most exciting biotechnology initiatives are now based in African countries. The Bill and Melinda Gates Foundation is putting substantial funding into these efforts – such as improved maize for poorer African soils, a project which is looking to get yield increases of 50% even where fertilizer is not available or the farmer cannot afford to buy it.

There's also the public-private partnership called Water Efficient Maize for Africa, using biotech to produce drought tolerant corn specifically for African smallholders facing the challenges of a changing climate. There's C4 rice, aiming to improve the photosynthetic capacity of rice and thereby dramatically increase yields.

Another Gates-funded project is based at the John Innes Centre in the UK and aims by 2017 to have cereal crops which fix their own nitrogen available for farmers in sub-Saharan Africa. The list goes on: there's biofortified cooking bananas in East Africa, and cassava fortified with iron, protein and vitamin A in Nigeria and elsewhere.

I haven't finished! There's resistance to cassava brown streak disease, which is currently spreading rapidly and threatens the staple crop for two out of every five people in sub-Saharan Africa.

And of course transgenic technology focused on conferring wheat rust resistance at the molecular level to head off the threat of a global pandemic which could otherwise threaten one of humanity's most important staple foods.

But if the activists have their way, none of these improved seeds will ever leave the laboratory. And this brings me, by way of conclusion, to the essentially authoritarian nature of the anti-GMO project.

All these activists, strikingly few of whom are themselves smallholder farmers in Africa or India, claim to know exactly which seeds developing country farmers should be allowed to plant. Those which are not ideologically approved by self-appointed campaigners should be banned forever.

The irony here is that predominantly left-wing activists, who are supposedly so concerned about corporate power, are determined to deny the right to choose to the most powerless people in the world – subsistence farmers in developing countries. In fact, this is more than an irony – it is a cruelty. And it is a cruelty which must stop, and stop now.

H.G. Wells is often quoted as saying that civilization is a race between education and catastrophe. The New Yorker writer Michael Specter, who wrote a great book about anti-science movements called 'Denialism', updates this, writing that civilization is a race between innovation and catastrophe.

This is surely no more true than today, when civilization is genuinely threatened by the twin catastrophes of climate change and ecological scarcity colliding with vastly greater food demand from a larger and wealthier population.

The solution is the same one that it always was – innovation – the uniquely human capacity to produce new tools which has saved our species so many times before from apparently inevitable Malthusian collapses. Therefore if we reject innovation now of all times we make catastrophe not just likely but probably inevitable.

This was indeed the warning the great Norman Borlaug left us with before he died. To quote:

"If the naysayers do manage to stop agricultural biotechnology, they might actually precipitate the famines and the crisis of global biodiversity they have been predicting for nearly 40 years."

In the final assessment the only way that conspiracy theories die is because more and more people begin to wake up to reality and reject them. Then perhaps there comes a tipping point where what was once received wisdom becomes increasingly understood for the foolish nonsense that it always was.

I think – I hope – that we are close to this tipping point today. And now, with just a little extra push, we can all join in consigning anti-GMO denialism to the dustbin of history where it belongs.

✂✎ MARK'S BIO

Mark Lynas is a British author, journalist, and environmentalist. The author of *The God Species: How the Planet Can Survive the Age of Humans* (2011), *High Tide: News from a warming world* (2004), and *Six Degrees: Our future on a hotter planet* (2007), which won the prestigious Royal Society Prize for Science Books. In 2009 he was appointed advisor on climate change to the President of the Maldives to be involved in the Maldives' effort to be the first carbon neutral country on Earth by 2020, until the President was despised in a military coup in 2012.

He is a frequent speaker around the world on climate change science, policy, and now on GM agriculture, and is also a Visiting Research Associate at Oxford University's School of Geography and the Environment.

- Mark Lynas

@Mark_Lynas

RANDOM THOUGHTS ON BIOTECH

BY FOURAT JANABI
Author

If one takes the basic premise that nature makes stuff better than we do — arguably the root of those who eschew GMO produce — and follow it through to its logical conclusion, we find something interesting. Starting at the beginning: some 3.5 to 3.8 billion years ago, there existed a single-celled replicator that was, most likely, the common ancestor of everything alive today. Now, if you are anti-GMO, harken back to the thought that recombinant DNA technology is unnatural. If that were indeed the case, then we could say with some confidence that we (humanity) wouldn't be here. The reason why is that nothing could've evolved from that original replicator. It would just be replicators ad infinitum, one after the undifferentiated other. Nothing would, or even could, change because random changes and mutations would not occur. (Even the original replicator would not have evolved to exist in the first place so we wouldn't have gotten that far.)

Food for thought: nature is the original engineer.

In order to go from that replicator to a 100-trillion celled human being, nature had to employ genomic engineering. The only difference between nature's style and our own is that nature's is directionless and purposeless – that is, there is no end goal in mind. Whatever happens, happens; good, bad, ugly, beautiful, painful, swift, agonizing, or any other of a hundred different combination.

Of all the species that ever existed, 99.9% are today extinct. Nature is not the benign process we think her to be, and though it is very easy to say that mother nature should be our guiding light (or spirit, or mother), there are 1.7 billion people who died of natural infectious diseases in the 20th century alone would not agree (if they could disagree, that is), neither, perhaps, would the 1.97 billion people who died of non-communicable diseases over the same time period. If we were to compare humanity's body count: all the wars, crime, subjugation, and intolerance against that of natures, we'd find that she more than trebled our own count, which horrifically stands at 980 million deaths. Be that as it may: it follows that we are here because of the natural process of genomic modification and there is nothing inherently unnatural in the process. Mutations happen: either nature makes them happen with no thought to the outcome, or we create a handful to suit our purposes with genetic engineering.

Genetic Modification in Nature.

Consider again the basic premise that nature makes stuff best and tack on another popular premise: that manmade is unnatural. Picture, from that first replicator onwards, nature haphazardly selecting for organisms. Preferentially selecting for those with beneficial mutations (better success in passing on their genes), selecting against those with detrimental mutations (less success), and being ambivalent towards those with benign mutations until, eventually, in the Rift Valley some few million years ago, a handful of primates left the trees, walked upright, and began evolving a larger frontal lobe, along with the spectacularly lucky coincidence of an opposable thumb. These concurrent lucky outcomes allowed their descendants to manipulate their environment with an ever-increasing degree of control as natural selection selected for finer motor control and intelligence over thousands and millions of years. Therefore, our intelligence and the manipulation of our environment are given to us by Mother Nature. Since every animal on this blue-green dot we call Earth uses to its advantage every trick and tool nature endowed it with − after all, those that don't often do not pass on their genes. It follows then, that, everything we do is the best possible way. We are made by nature, therefore everything we do is natural and, therefore, everything we are doing now is the best possible solution because it is natural. As you can see, this line of reasoning (natural is better than human-made) is a slippery slope and is, plain and simply, ill defined.

The distinction between nature, human culture and technology is an arbitrary distinction. We do the things that we do now because of our naturally endowed capacity. But, perhaps another way to put it is this: after 3.8 billion years, Homo sapien sapiens evolved to mold it's own evolvability (technology) continuing the process of selection, in the process superseding, in some small domains, natural selection. We are the first species that does not live entirely within the constraints of natural selection. That does not mean we don't live in a selection process − just that we override nature's and institute our own. In time, we rely less on natural selection and more on environments of our own choosing − but it is so because nature made it so. Ants make anthills, beavers make dams, birds make nests, and we make technology; practical, virtual, bio, and it's all natural.

Selection.

Evolution happens regardless of whether we rework it to our advantage (biotech crops) or leave nature be. We may then break down the various categories as such:

- Evolution is natural selection by random mutation

- Pre-Industrial (i.e., organic) agriculture is *artificial* selection by random mutation

- 20th century (conventional & organic) agriculture is *artificial* selection by *accelerated* random mutation (mutagenesis)

- GM agriculture is *artificial* selection by *purposeful* mutation

As you can see above the changes are changes in degree, not kind. To label one unnatural is to label them all unnatural; as Richard Dawkins once said, "no agriculture is natural." To be pedantic, the natural way of life for a human being is the hunter-gather lifestyle. Humans have been around for approximately 200,000 years, and agriculture has been practiced for only 10,000 years. Each of the above labels is evolution (that is to say, natural), continued; it exists on a spectrum. Something has to fulfill both the selection (choosing) process and the random (mutation) process in evolution. If we are happy to leave it at organic, then nature, which has neither direction nor purpose — and evidenced by her 3.67 billion person death toll in the 20th century alone from just two categories of death — has neither our health or longevity in mind; or we fulfill the selection process, which, paradoxically to those who hold nature in high esteem, it has given us the ability to so (albeit, randomly).

While the result of recombinant DNA technology may be labeled unnatural — because it doesn't exist in nature, not because it can't — the same cannot be said of the technology that produces such food. We are co-opting nature's methods to make food, not playing God. Some may dispute that fact by saying that a fish gene could never wind up in a tomato, but that would be betraying a fundamental concept of evolution: nature uses the same genes over and over again in all manner of disparate, sometimes far-removed, creatures. There are no such thing as fish genes, tomato genes, or human genes; there are only genes that perform specific functions and that operate according to the principles of natural selection. Take an example: your genome is the combined genome four times over of the amphioxus fish-like marine <u>chordate.</u> The marine chordate's genome, which is a 1-cm little fish that still exists today, has in the course of Earth history, mistakenly copied over on itself twice and those two mistakes resulted in every land animal today, and you. If nature can turn a little fish into you (and its genome is still inside you), then why is it so distasteful that we put disparate genes where we need them? Uncertainty may be the first thing that comes to mind, but nature had no idea what she was doing either. After all, 99.9% of all species died out for us to be here. So it's not like *she* knew what she was doing. We, however, while lacking complete certainty, do have a fair amount of knowledge on how all this works, genomic modification has been ongoing since the 1970s. It wasn't discovered yesterday, in which case, I, too, would share some concern. (In fact, the first moratorium on genetic technology was instituted by the scientists involved in the late 70s/early 80s to work out the myriad ethical concerns — everyone else were late to the party by a few decades; most of the concerns were, and most of the work put in hiatus, to hash out the problems raised.)

What's The Point?

There is a movement to demonize GM technology, even conventional agriculture, and then, for some reason, a concomitant wish to return to a mythical agricultural past. Organic agriculture is fine, there's nothing necessarily wrong with it, but we can't feed the world with it. Paul R. Ehrlich's book *The Population Bomb* stated in 1968 that in the 70s and 80s, mass famines would ensue as we wouldn't be able to make enough food, and any efforts to avert such a disaster are a waste of time and should be scrapped. (Thomas Malthus said the same thing in 1798.) Ehrlich wrote that "the battle to feed all of humanity is over. In the 1970s the world will undergo famines — hundreds of millions of people are going to starve to death in spite of any crash programs embarked upon now." Why didn't the predictions of mass starvation and disaster come to pass? Well, they would have if we listened to him and did nothing. Luckily we didn't. Instead, we developed the technologies that allowed us to increase yield to a stupendous degree to avert such a disaster. That is, we bid goodbye to organic agriculture.

Context is Key.

Since 1961, we've increased yield by 300%, and only had to increase our land use by 12% to do so. How? We used technology to drastically increase yield and avert the predicted disaster of Ehrlich and many others. Said differently, if we kept farming organically, mass famine would have ensued. Without such yield increases (thanks to plant science), we would have had to use two Latin America's of arable land to compensate. Or, more likely, the predicted mass starvation would have occurred.

If in the 1960s, when the world population was less than 3 billion people, the propagation of organic farming as the sole agricultural method would have resulted in disaster, how could it possibly help us now when we are 7 billion people and on the way to 9-10 billion people? The majority of that increase in yield has come from plain conventional agriculture and plant breeding, but now our yields are coming up against a glass wall for that type of plant science, and GM foods are the next process to take us forward to surmount the coming set of problems (namely: a 70% increase in food production while also decreasing land requirements). And, while we still have a starving billion today, it is not because we can't create the food, but we can't get it to them. The solution to world hunger is for those most afflicted by it to be able to grow their own food, instead of relying on food aid and handouts as if band aids were being applied to a broken bone. Organic farming will not suffice for Sub-Saharan Africa; they need heat-tolerant and drought-resistant strains. They already don't have any biotechnology or conventional agriculture; ergo, organic farming, which is what remains, has failed them.

Future Problems.

By 2050, we will need to almost double yield without an increase in land usage — in fact we'll need to decrease land usage as agriculture is one of the biggest contributors to climate change. We will not accomplish this by going back to low-input agriculture — though it won't go anywhere for those who still want it. I've made the case before that Vertical Farming (VF) could do the trick. VF certainly is capable. We could grow food in cities at a productive rate 5-10 times that of horizontal farming, use no pesticides, and with vastly reduced water requirements, but what if the mass migration from horizontal farming to vertical farming never takes place? The technology was invented in the 1950s by the US military and then nobody did much of anything with it for 60 years since. What if that no-usage scenario repeats itself? We cannot afford to stand idly by and hope that everything will go according to plan. We need contingencies and redundancy instead of wishful thinking and vague plans. GM agriculture can significantly contribute on that count. We have been growing and eating GM food for almost 20 years; in that time, we've spared the environment 474 million kilograms of pesticide use. (Don't forget, organic farming uses pesticides too, and organic pesticides aren't automatically better for the environment.) In 2010, 23 billion kilograms (50 billion lb.) of CO_2 was not released into the atmosphere because of GM technology (the equivalent of 10.2 million cars removed from the roads for a year). In 2011 51% of the economic benefits of GM seeds ($19.8 billion) went directly to farmers in developing countries helping them rise up out of subsistence farming and poverty, and since 1996, about 50% of the $98.2 billion productivity surplus from GM crops have gone directly to farmers in both developed and developing countries. On the health side side, in America, the country that eats the most GM food, cancer mortality over the last twenty years are down 20%, so the promised health apocalypse that many have warned about were coming have not materialized — moreover, those that have materialized have explanations, and those explanations are in no way related, directly or indirectly, to GM food. No new allergens have been identified, nor have any biological mechanisms of harm been identified by public scientists.

Potential Benefits.

Recently, we passed peak farmland, which unlike peak oil or peak water actually has positive connotations for humanity, and especially for the environment.

Since 1961, to get us the aforementioned 300% yield increase required to stave off worldwide famine, we spared the equivalent landmass of the USA, Canada, and China. Try to imagine the destruction of forestry that that would have entailed, if we had rioted then against plant science as we do now. To be an environmentalist is, by definition, to support the conservation of nature. To support the conservation of nature should be, by definition, to support conventional agriculture as it uses less land to grow that food — going forward, this will entail supporting the cultivation of GMOs, there's no two ways about it. PG Economics noted that if, in 2010, those biotech crops already available were removed from the market, to keep production steady farmers would have had to plant an additional 5.1 million ha of soybeans, 5.6 million ha of corn, 3 million ha of cotton, and 0.35 million ha of canola. Equivalent to an additional 8.6% of arable land in the US. Yet, this is what activists would have us do, remove all GM crops, necessitating the further destruction of forestry and nature for human purposes.

If we continue on our current path of increasing yields using science and biotechnology, the authors of the Peak Farmland study conservatively estimate that we could return 146 million hectares to nature by 2060, with high estimates at 400 million hectares (roughly double the area of the USA, east of the Mississippi). The coming generation of biotech crops, many of which will have significantly reduced pesticide use (some using no pesticides at all), fix their own nitrogen (reducing river pollution), increased nutrition along with many other benefits reducing malnutrition and disease. In other words, the potential to alleviate many of the environmental ills associated with conventional farming, but many such seeds are locked away due to the intense furor to GMO use, allowing only those few seeds through that the giants can afford to push through the regulatory burden, which will return a profit.

So, as we move forward into the future, we'll give back hundreds of millions of hectares of farmland to nature, and if we move forward with biotechnology, we'll give back even more. How can any environmentalist organization like Greenpeace and Friends of the Earth think that a bad thing?

Big Ag.

Are there problems, real problems, with biotechnology that have been covered or up concealed? With the technology, we find no problems that aren't present in other forms of agriculture. As the National Academy of Science, and many prestigious scientific organizations concluded, the process itself is no more inherently risky than any other method; it is simply a refinement of previous methods, far more precise and <u>quicker.</u> Biotech crops usually have between 1 and 3 genes altered, but every new generation of organic and conventional crops that reproduces sexually will have a few different genes in there too (this is why farmers buy seed, and that seed-purchasing predates Monsanto). (They are inevitable: a DNA copying error, a passing cosmic ray etc., will, and do, induce genetic mutations. To say there is uncertainty in GMOs is likewise to admitting that there is uncertainty in any new generation of plant or animal. The average human offspring carries about 100-200 mutations, but they are still people. Food with 1-3 added genes is still food.)

On the business side is where we find many that many folks have *a priori* problems. But these problems are indicative, and suggest the need of, business reform, patent reform, and competition. Not the outright banning of the technology (which is just not possible, anyway). These *a priori* (business) problems have somehow co-mutated into advocacy against GMOs in general instead of where it should actually be directed, lack of competition. But that lack of competition is due to the overbearing regulatory burden on GM crops. The problem was, is, and will be for some time, that the lack of competition has been institutionalized due to the initial, fierce, and hysterical advocacy by the anti-GMO activists; and round and round the circle we go, as the increased advocacy only exacerbates the problems activists think they are trying to stop. If I was Monsanto, I'd be funding the antis for the monopoly it brings. The intense backlash against biotechnology has only cemented the power of those few who first began exploring the field, as only they have been able to afford it, thus far. Even then, the scale of abuse, often leveled at Monsanto, rivals the misinformation that the Catholic Church spouts against condom use on the continent most ravaged by aids, likening condom use to be a greater danger than the ravages of aids.

We need to stop pretending that only Big Ag and Monsanto lobbies. The organic movement as a whole spends $2.5 billion a year on advocacy and lobbying. (Big Organic?) We need to stop thinking that Monsanto is after world domination: the global GM seed market in 2012 was $14 billion (world domination with only 0.0002% of global purchasing power? If they can pull that off, they probably deserve it), while organic food sales were $60 billion worldwide. (The total value of those GM crops when harvested was around $65 billion at that time.) We need to know that all farms strive to use the least amount of pesticides required, as it is their biggest expense, and that synthetic chemicals are not *a priori* worse than organic chemicals — in fact, quite often, it's the opposite. In other words, we need to with the facts as they are, not as we want them to be.

For whatever problems we have today, the solution is not to ban it, nor proffer simpleton solutions, but to weigh the risks vs. the rewards and act appropriately. One can only do that with evidence, i.e., science. The solution is to study, to research, and to have reasoned debates among experts on the pros and cons; but above all, keeping in mind the effects on people far and wide around the world. Food security and a heavy disease burden (usually going together) undermine society at every level of functioning. To fix them is to advance significantly in all other matters of societal dysfunction. Who knows how many Newtons, Einsteins, and Curies we are losing to lack of food, clean water, and education every year while we bicker over functionally equivalent types of food. The consequences of this debate have far-reaching consequences worldwide. If people don't want to eat GM food, they don't have to, but to stop others from making their own choice is to deny them a choice. The liberal movement in America and Europe is pro-choice when it comes to matters of female reproduction — and rightfully so! Yet, move the topic to food, and they swiftly change to being anti-choice, even though the ramifications for billions of poor people around the world are far worse than for a women in a forced pro-life environment. The common refrain is we have a right to know, where are the labels? But, there is no need for a label, the opposite label — "Certified Organic" — already means GM-free. Why must society bifurcate the food system even further?

Instead of focusing on legitimate problems with the business, competitive, and legal environment, red herrings are thrown this way and that: that organic food is nutritionally superior; a meta-analysis covering 162 studies over a 50-year period says their not, and any nutritional differences are unlikely to have a significant outcome on health, anyway. *Facts* are thrown out stating that organic is environmentally superior to all other forms of farming, despite the fact the answer is far more nuanced, and errs more to the side of no. We are told that farmers are using GMOs to lather their fields in Roundup, yet the National Academy of Science wrote, "When adopting GE herbicide-resistant (HR) crops, farmers mainly substituted the herbicide glyphosate for more toxic herbicides." (A report from the National Research Council even gave an impressive list of GM benefits including: improved soil quality, reduced erosion and reduced insecticide use, but everyone focused instead on the little nuggets of bad news instead of the truckload of good news.) In using GMOs we use less toxic pesticides, and the result is a net environmental benefit. Instead of learning about real yields on GMO, we get the 'Union of Concerned Scientists' telling us that 'intrinsic yields' haven't increased since the inception of GMO, even though intrinsic yield tells you nothing, but total yield really has increased, significantly. But the most destructive effect of this headline-grabbing debate fiasco is as Pamela Ronald, professor of plant pathology at the University of California wrote, "As it now stands, opposition to genetic engineering has driven the technology further into the hands of a few seed companies that can afford it, further encouraging their monopolistic tendencies while leaving it out of reach for those that want to use it for crops with low (or no) profit margins."

Red herrings are red for a reason, they are meant to distract, instead of inform, you. We need some green herrings. Science has provided them, almost everyone has ignored them, and those who tout them are labeled "shills". We need to stop seeing the world with ideological filters.

Choices.

Those of us with the ability to read this book have the luxury of choice when it comes to choosing between organic and conventional / GM agriculture. But more than 900 million people will go to bed hungry every night (16 million people of whom will die of hunger this year) without that luxury. Half the planet's population remains malnourished, then there are the one to two million people (670,000 of whom are under five years of age) who'll die from Vitamin A deficiency this year. These folks, and more, need more nutrient dense food to not only survive, but prosper. Those 1-2 million folks who die of Vitamin A deficient will not, in point of fact, be thankful to Greenpeace for their 13-year blockade of GM Golden Rice that could save them, they'll die slow, painful deaths instead, only to be replaced by more kids, many of whom will die too. To fix that problem —

which is not only a moral necessity — also reduces the burdens of population growth and resource management.

It's time we got out of our *First World* bubble. There are many difficult decisions out there that we will need to make in the future. The issue of GM food is not be one of them.

There is, despite the hysteria, a scientific consensus on the safety and risk profile of GM technology. Almost every scientific organization, from the National Academy of Sciences to the Royal Society, has evaluated the evidence and come to that conclusion. There are over 650 peer-reviewed studies to back up the claim; one-third of which are independently funded. Aside from a few deniers, we trust our scientists on climate change, don't we? They are shouting from the rooftops about the dangers of climate change, and how little time we have left to reverse course. You'd think if there were a comparable danger from biotech, you'd have more than a handful of scientists speaking up. So, why don't we trust them on biotech?

Ingo Potrukus, inventor of the Golden Rice that was to be given patent-free to the developing world had something quite damning to say of the food movement that is demonizing GMOs: "If our society will not be able to 'de-demonize' transgenic technology soon, history will hold it responsible for [the] death and suffering of millions: people in the poor world, not in overfed and privileged Europe, the home of the anti-GMO hysteria."

It does no good to deal in hypotheticals such as: if we wasted less food, there'd be enough for everyone (you wouldn't be able to ship it to them); if more people were charitable, everyone would be ok; if we switched to organic agriculture, we could feed everyone (wrong), along with many others. Despite the fact that many of them are wrong or idealistic, they presume people being rational, informed, and having access to and accepting unadulterated and uncensored good, reliable information in context. Is that likely to happen anytime soon? The cries of the anti-vaccacinists are still putting kids (and society at large) in danger; the chant of the climate-deniers only delays needed progress; but on issues of food security, arguably the most important of all, everyone will see reason?

Changing People or Inventing Technology.

Perhaps an easier way to put it is this: is it easier to change the hearts and minds of billions of people with all their complexities and interrelationships, or is it easier to invent new technologies that solve the issues for those affected without bothering or relying on strangers in the West? The climate movement has struggled to change the hearts and minds of people and politicians for over twenty years and we've got very little to show for it. Let's not continue making the same mistake with food. Changing the consumption habits of one billion westerners – if that is even possible – will take a long time, with no certainty of success. Meanwhile, half the planet's population suffers from malnutrition, and 1/7 from hunger in food insecure areas. The technologies to feed them using less land, cheaper inputs, and more nutrients are here and now. They are safe, capable, and predictable, regardless of how shrill the opposition to them is from well-fed oppositionists who've never felt the sensation of hunger. It's time to deal with the facts, but above all, it is time to value human lives above ideology. The intentions and hearts of the bored, guilted sensibilities of Western activists who grumble at a skipped lunch is in the right place; their proposed solutions and flawed reasoning are not only wrong, but are having precisely the opposite effects.

They are plenty of problems in agriculture. The vehement backlash against biotechnology is distracting from discussing, debating, resolving, and funding those problems. Biotechnology won't solve every problem, but they will help substantially. In fact, the co-use of biotech crops alongside organic crops — in what is called a refuge zone — significantly curtail pest resistance allowing us to get the most of our <u>crops.</u> It may be that the bright agricultural future within our grasp uses both systems side by side, as argued earlier by Ramez Naam.

We need to realize that feeding 7 billion, let alone 9 to 10 billion people in the near future, isn't going to be easy. If it fits on a Facebook photo as a caption, you can rest assured it will solve nothing. This chapter is 5,000 words long and is barely scratching the surface. This book is almost 40,000 words and barely digs beyond the tip of the iceberg. Some silly shared photo on Facebook demonizing Monsanto, chemical use, glyphosate, or showing a tomato with a hypodermic needle in it not only shows you things out of context, they detract from the conversations we should be having, if it isn't an outright lie to begin with (hypodermic needles are not used in the genetic modification of... well...anything).

No one, least of all the scientists and farmers contributing to food security, is pretending that the current state of agriculture is perfect. But it is far and away more advanced than the ancient and recent past, and any and all solutions to today's problems will only come from the application of more research, science, evidence, and technology.

With that, I duly hope that you've picked up something from this book, and I especially hope that that something includes peace of mind. The amount of fear, paranoia, and hysteria one finds on the Internet is as disconnected empirically from reality as astrology is to astronomy and alchemy to chemistry. Carl Sagan, in his usual candor, said it best: "Modern science has been a voyage into the unknown, with a lesson in humility in every stop. Our common sense intuitions can be mistaken, our preferences don't count, we do not live in a privileged reference frame."

Whatever your impression of genetically modified organisms may be, if we do not ground the discussion in the strictest of facts, we stand to lose very much. GMOs might not solve everything, but to not use every means at our disposal to combat climate change, food-security (and by extension, poverty), water use, and shelf-life for ideological reasons is inviting fragility and vulnerability to our food system.

෴

FOURAT'S BIO

Fourat Janabi is a photographer, blogger and author. of *Random Rationality* and *S3: Science, Statistics and Skepticism*. He blogs at RandomRationality.com.

- *Fourat Janabi*
@fouratj

FACTS

1. Crop biotechnology has been the fastest adopted innovation in agricultural history growing from 1.7m Ha in 1996 to 170m Ha in 2012

2. In 2012, the global value of the GM seed market was $14.84 billion

3. The adoptions of biotech-enhanced crops reduce chemical pesticide use by 37% and increase farmer profits by 68%

4. The economic benefits of GM crops amount to an average of $117 per hectare, with the majority accruing to farmers in developing countries

5. Farmers, on average, receive about $3.33 back per dollar invested in GM seeds

6. At least three-trillion meals containing genetically modified ingredients have been consumed

7. There are 650 peer-reviewed scientific studies that show no harm from crop biotechnology; of those, 1/3 are independently financed (see genera.biofortified.org for full list)

8. Biotech crops have yields, on average, 7-22% higher than conventional farming, which yields, on average, 20-35% more than organic farming

9. In China, GM crops have increased yields by 7%

10. In Australia and China, pesticide use was reduced by 80% on GM crops

11. Biotechnology reduced pesticide spraying between 1996-2012 by 503 million kgs (approximately 1.1 billion lbs.), reducing the Environmental Impact Quotient (EIQ) of farming by 18.3%

12. Crop biotechnology has grown an additional 311.8 million tons more food in the last 15 years

13. Crop biotechnology has saved 109 million Ha of land from being ploughed

14. Net agricultural CO2 savings related to Biotechnology as related to climate change are:

    1. From 1996 – 2007, biotech reduced CO2 emissions by approximately 10.5 billion kilograms (23 billion lbs.)

    2. In 2012, atmospheric CO2 emissions were reduced by 27 billion kilograms (59 billion lbs.)

CONCLUSION

National Academy of Science

All evidence evaluated to date indicates that unexpected and unintended compositional changes arise with all forms of genetic modification, including genetic engineering. Whether such compositional changes result in unintended health effects is dependent upon the nature of the substances altered and the biological consequences of the compounds. To date, no adverse health effects attributed to genetic engineering have been documented in the human population.

~

European Commission

The main conclusion to be drawn from the efforts of more than 130 research projects, covering a period of more than 25 years of research, and involving more than 500 independent research groups, is that biotechnology, and in particular GMOs, are not per se more risky than e.g. conventional plant breeding technologies.

Royal Society of Medicine

Foods derived from GM crops have been consumed by hundreds of millions of people across the world for more than 15 years, with no reported ill effects (or legal cases related to human health), despite many of the consumers coming from that most litigious of countries, the USA.

~

World Health Organization

GM foods currently available on the international market have passed risk assessments and are not likely to present risks for human health. In addition, no effects on human health have been shown as a result of the consumption of such foods by the general population in the countries where they have been approved.

~

The International Council for Science

Currently available genetically modified foods — and foods derived from them — have been judged safe to eat, and the methods used to test them have been deemed appropriate.

~

The American Association for the Advancement of Science

The science is quite clear: crop improvement by the modern molecular techniques of biotechnology is safe.

~

American Medical Association

There is no scientific justification for special labeling of genetically modified foods, as a class, and that voluntary labeling is without value unless it is accompanied by focused consumer education.

~

The American Society for Cell Biology

Far from presenting a threat to the public health, GM crops in many cases improve it.

~

The Union of German Academies of Sciences and Humanities

In consuming food derived from GM plants approved in the EU and in the USA, the risk is in no way higher than in the consumption of food from conventionally grown plants. On the contrary, in some cases food from GM plants appears to be superior in respect to health.

~

The French Academy of Science

All criticism against GMOs can be largely rejected on strictly scientific criteria.

~

The American Society for Microbiology

The ASM is not aware of any acceptable evidence that food produce with biotechnology and subject to FDA oversight consti-
tutes high-risk or is unsafe. We are sufficiently convinced to as-
sure the public that plant varieties and products created with
biotechnology have the potential of improved nutrition, better
taste and longer shelf-life.

~

The International Society of African Scientists

Africa and the Caribbean cannot afford to be left further behind
in acquiring the uses and benefits of this new agricultural revo-
lution.

~~~

*Our species needs, and deserves, a citizenry with minds wide awake
and a basic understanding of how the world works.* -- Carl Sagan

The End